HISTORY OF
SCIENCE · AND ·
TECHNOLOGY
REPRINT SERIES

The Gene

A Critical History

Elof Axel Carlson

IOWA STATE UNIVERSITY PRESS ■ Ames

Elof Axel Carlson is Distinguished Teaching Professor in the Department of Biochemistry, State University of New York, Stony Brook.

Originally published by W. B. Saunders Company © 1966 by W. B. Saunders Company. All rights subsequently assigned to Elof Axel Carlson.

This edition published in 1989 by Iowa State University Press, Ames, Iowa 50010

Text reprinted from the original without correction

History of Science and Technology Reprint Series

First printing, 1989

Library of Congress Cataloging-in-Publication Data

Carlson, Elof Axel.
    The gene: a critical history/Elof Axel Carlson.
        p.   cm.—(History of science and technology reprint series)
    Reprint. Originally published: Philadelphia: W. B. Saunders Co., © 1966.
    ISBN 0–8138–1406–5
    1. Genetics—History. I. Title. II. Series.
QH428.C246   1989
575.1'09—dc20   89–24427

Dedicated
to

MORRIS GABRIEL COHEN

Who introduced me to the art of reading

PREFACE

My interest in writing a book on the gene concept began during the preparation of a review article on complex loci (*Quarterly Review of Biology*, March 1959). At that time I surveyed a major portion of the work written on gene structure and I was struck by the numerous instances of independent discovery, periods of obscurity, and spurious philosophic attitudes that subsisted underneath the apparently smooth transition of ideas and experimental progress that reviews and texts alike tend to produce. I did not feel, at that time, that I could do justice to the evaluation of the creative process in the development of the structure of the gene. I chose, instead, to prepare a conventional review of the results, rather than the causes, of the investigations on complex loci. A sabbatical leave from my normal University duties has now given me the solitude to cover the historical aspects of this problem.

The result of my interest in this topic is *The Gene: A Critical History*. It is intended for two major groups of readers—students of genetics and students of the history of ideas or of the history and philosophy of science. In planning this book, I decided to approach the problem in an unconventional way. Ordinarily a history tends to be a narrative in which chronology and fact become highly distilled. I have chosen instead to restrict the history of the gene concept to a series of themes, each theme based on one unique or dominant concept. Frequently these concepts were received with hostility and their success or failure required a number of years. In describing such conflicts I have tried to explore the reasons why each concept emerged in its original form, what the experimental approaches were that made it a strong or weak idea, and most important, what the underlying philosophy or point of view was which motivated each contender in the conflict. To this extent, these chapters are not strict history; they are artificial in a contrived way. To give a flavor of the debates, I have avoided paraphrase; quotations from the original works are used to convey the styles of the participants. This gives to many of these chapters a theatrical or dramatic quality. They are genetic dialogues. This adds to the readability of many topics that might be tedious to the uninvolved reader, but the objectivity lost by the "excision" process must be kept in mind. The historical and philosophical implications of this study of the gene concept are treated at length in the last two chapters. So, too, is my own evaluation of the significance of this concept and its status today.

The Gene: A Critical History is not a comprehensive review. To attempt so vast a scope I would have had to construct a series of volumes, probably three or four times the length of this book, with more than half of it devoted to the years from 1955 to the present. The number of papers on gene structure and function in recent years is too large for a representative historical evaluation; nor is the flow of ideas and information likely to diminish in the near future. Perhaps a perspective will be obtained by geneticists and historians in the next decade. I have not been able to resolve this problem and my choice of material in the molecular fields is more likely to be biased than is that for the earlier non-molecular fields.

The student of the history of science usually deals with personalities and events many generations removed. He has the security that his interpretations will offend only other historians. This is not the case for a history of the gene concept. Many, if not most, of the contributors to the gene concept are alive today. This unique circumstance originally tempted me to consider visiting most of these participants and collecting a large number of taped interviews from which I would be able to abstract patterns or conclusions on the development of different ideas. I rejected this approach, in favor of the one adopted here, because I did not want to come to a conclusion on my own which one of my colleagues would interpret as a betrayal of confidence or an insult to his conversation. Finally, I rejected it out of fairness to the dead. They do not have the privilege of advising me that I am misinterpreting their intentions or their results. I do believe that the interview approach should eventually be used. For the purpose of this book, however, I have chosen the library as the ultimate source of my information. The history of the gene concept will be a lifetime hobby for me; I would be particularly grateful for corrections and new sources of information, particularly original correspondence, diaries, and unpublished manuscripts.

As a minor contributor to the gene concept, I have had the advantage of participating in genetic research; this has been very valuable in interpreting the effectiveness of design in those experiments which determined the survival of ideas. I also wish to acknowledge the influence of the graduate coursework taught at Indiana University when I was a student there some ten years ago. Standing out among all these academic influences was a course, "Mutation and the Gene," taught by H. J. Muller in 1955 as a series of forty lectures. While Muller's approach and my own are different, I have been strongly stimulated by the imaginative, creative, detailed, and thrilling panorama of genetics which Muller "relived" in the process of giving his course. For this "contagion" of ideas I am grateful. However, it is my views, and not Muller's, that should be questioned if the reader feels that an interpretation seems unfair.

The quotations used in The Gene: A Critical History retain their original style, syntax, and spelling. All italics within quotations are the original author's. Occasionally, I have added a word in parentheses. The number following a quotation or a cited article corresponds to the number of the article in the References.

The first draft of the manuscript for The Gene: A Critical History was begun in February 1965 and completed in May 1965 at Woods Hole, Massachusetts. The manuscript was revised in July and August 1965 in Rochester, Indiana, and returned to the publisher, with illustrations, in the fall of that year. While the book was in production I obtained copies of two books on the history of genetics which may be of interest to the reader:

A. H. Sturtevant (1965): *A History of Genetics*, Harper and Row, N. Y., 165 pages.

L. C. Dunn (1965): *A Short History of Genetics*, McGraw-Hill, Inc., N. Y., 261 pages.

In addition, I was sent a copy of an article by H. J. Muller which is of special interest. The article, "The Gene Material as the Initiator and Organizing Basis of Life" presents Muller's own evaluation of his contributions to the gene concept, particularly in relation to evolution and present day (molecular) views. Muller's article will be published in the semi-centennial (1965) anniversary issue of the *American Naturalist*.

I wish to thank the University of California, Los Angeles, for granting me a sabbatical leave to prepare this book. The reading for this work was made almost effortless by the magnificent facilities of the Lillie library at the Marine Biological Laboratory, Woods Hole. The winter solitude for reading, writing, and thinking was supplemented by the unique, extensive reprint collection housed there. I urge my colleagues who have not done so to send reprints to this library for the use of future generations of students and scholars.

I wish to thank my wife, Nedra, for her encouragement and patience while this book was being written and for the many hours she spent in typing the manuscript. I also thank my brother, Roland, who prepared the illustrations, and Mrs. Florence Okuyama, who typed much of the final draft. The criticisms of J. L. Southin and Marcus Rhoades, who read the entire manuscript, were particularly useful. I alone, however, accept sole responsibility for any errors which appear.

Los Angeles   ELOF AXEL CARLSON

Preface to the Reprint

After almost 25 years both the history of genetics and the gene concept have undergone considerable development. The gene today is comprehensible to biologists only through biochemistry. In eukaryotic cells the gene is organized into informational segments (exons) with intervening sequences (introns), which are often considerably larger than the exons. Some genes have numerous exons, others few. They probably are of immense evolutionary significance in the formation of large proteins assembled from the processed RNA that is enzymatically trimmed of its introns. The second major feature of the gene that

should be added is the technical development that permits rapid se-
quencing of the nucleotides and the cutting of DNA into pieces at
specific sites through restriction enzymes. The recombinant DNA technology using these techniques has resulted in both applied and basic
scientific contributions. None of these dramatic and overwhelmingly
important findings has changed the basic idea of the gene as the unit of
inheritance that is transmitted in Mendelian fashion according to the
meiotic distribution of chromosomes.

The history of genetics has been enriched by many approaches.
Formal biographies have appeared of Muller (Carlson, *Genes, Radia-
tion, and Society: The Life and Work of H. J. Muller,* Cornell University
Press, 1981); Morgan (G. Allen, *Thomas Hunt Morgan: The Man and
His Science,* Princeton University Press, 1978); McClintock (Evelyn Fox
Keller, *A Feeling for the Organism: The Life and Work of Barbara
McClintock,* Freeman, 1983); and Wright (W. B. Provine, *Sewall
Wright and Evolutionary Biology,* University of Chicago Press, 1986).
Autobiographies are also numerous, including those by Watson (*The
Double Helix,* Atheneum, 1967); Jacob (*The Statue Within,* Basic
Books, 1987); Luria (*A Slot Machine, A Broken Test Tube,* Basic Books,
1985); and Crick (*What Mad Pursuit,* Basic Books, 1988), although
these tend to focus more on personality development than on an appraisal of a career. Many new approaches have been tried in present
genetics from a historical perspective. Olby attempted to trace the disci-
plinary influences leading to Watson and Crick's 1953 work (The Path
to the Double Helix, University of Washington Press, 1974); Judson
used the interview approach to trace the ideas in the history of molecular genetics (*The Eighth Day of Creation,* Simon and Schuster, 1979);
Sayres (*Rosalind Franklin and DNA,* Norton, 1975) used the biases
against women as an attempt to understand Rosalind Franklin's career;
Sapp (*Beyond the Gene: Cytoplasmic Inheritance and the Struggle for
Authority in Genetics,* Oxford, 1987) used a sociological model of
power struggles in interpreting the fate of cytoplasmic inheritance; and
Dronamraju (*The Foundations of Human Genetics,* Thomas, 1989)
used a philosophic analysis of Kuhn's model of scientific revolutions to
interpret human genetics as a Kuhnian "revolution."

A thorough discussion of these new trends in the gene concept and
the history of genetics would require another book. This reprint is
intended to acquaint a newer generation of scientists, philosophers, and
historians with the earlier history of the gene concept in its classical and
early molecular phase. Those desiring further insight into the history of
genetics will benefit from a reading of the works cited in this preface.

ELOF AXEL CARLSON, 1989

CONTENTS

ACKNOWLEDGMENTS

Permission to quote was generously given by the following publishers and journals:

ADVANCES IN ENZYMOLOGY: 259
ADVANCES IN VIRUS RESEARCH (ACADEMIC PRESS): 289
AMERICAN NATURALIST: 3, 33, 51, 68, 69, 70, 73, 105, 106, 118, 119, 141, 145, 161, 187, 198, 205, 207, 218, 220, 223, 225, 231, 237, 242, 265, 276, 285, 295, 298, 310
AMERICAN SCIENTIST: 286
CAMBRIDGE UNIVERSITY PRESS: 12, 18, 24, 177, 273
COLD SPRING HARBOR SYMPOSIA: 2, 61, 85, 102, 136, 171, 184, 191, 261, 269, 283, 293
COLUMBIA UNIVERSITY PRESS: 262
EXPERIENTIA: 150, 189
GENETICS: 32, 52, 55, 58, 98, 115, 122, 124, 125, 127, 157, 181, 182, 183, 222, 247, 267, 300, 301, 303, 307
HARVEY LECTURES (ACADEMIC PRESS): 30
JOHNS HOPKINS PRESS: 40, 179, 252
JOURNAL OF EXPERIMENTAL ZOOLOGY (WISTAR INSTITUTE): 19, 49, 53, 154, 213, 216, 219, 224, 322
JOURNAL OF GENETICS: 23, 25, 117, 138, 232, 241, 274, 275, 311
JOURNAL OF HEREDITY: 72, 101, 134, 211, 235
JOURNAL OF MOLECULAR BIOLOGY (ACADEMIC PRESS): 91, 170, 176, 193
MCGRAW-HILL BOOK CO.: 244
NATURE: 37, 47, 60, 94, 128, 130, 132, 135, 146, 168, 251, 263, 270, 288, 315, 316
PERSPECTIVES IN VIROLOGY (BURGESS PUBLISHING CO.): 290
PROCEEDINGS OF THE NATIONAL ACADEMY OF SCIENCES: 4, 31, 34, 38, 41, 42, 45, 62, 75, 76, 77, 78, 81, 107, 137, 151, 155, 160, 163, 178, 180, 185, 190, 195, 221, 229, 249, 253, 254, 256, 264, 268, 279, 281, 282, 306, 312, 314, 320, 321
PROCEEDINGS OF THE ROYAL SOCIETY: 11, 92, 131, 209, 243, 260
SCIENCE: 22, 50, 54, 65, 71, 82, 87, 114, 140, 153, 156, 162, 199, 200, 201, 202, 203, 210, 228, 238, 257, 266, 277, 278, 291, 294, 296, 305
SYMPOSIA OF THE SOCIETY FOR EXPERIMENTAL BIOLOGY (ACADEMIC PRESS): 89
VIROLOGY (ACADEMIC PRESS): 313

The numbers which appear after the names of journals or publishers above refer to citations in the bibliography on page 273. The appropriate author, book, editor, publisher, journal, volume, year, title and pagination are listed in these bibliographical references. I am also grateful to my colleagues, too numerous to mention, who gave permission to quote from their works.

CHAPTER 1

A Modest Discovery

Some discoveries are announced under modest circumstances, are slow to gain recognition, and may await independent rediscovery before their significance is realized. This was the fate of Mendel's experiments with the garden pea. By contrast, Darwin's theory of Natural Selection was announced to an anxiously awaiting scientific audience, its fame was immediate, and a deluge of commentaries and investigations was stimulated by it. Perhaps, in 1902, William Bateson was moved by the historical accounts of Darwin's success when he wrote that "an exact determination of the laws of heredity will probably work more change in man's outlook on the world, and in his power over nature, than any other advance in natural knowledge that can be clearly foreseen."[10] Bateson was bringing to the British public the rediscovery of Mendel's theory of heredity. The rediscovery was made two years earlier by three European botanists of whom Hugo de Vries was the most prominent. It was de Vries's account of the confirmation of Mendel's experiments and theory which first attracted Bateson's attention and stirred his imagination: "In the year 1900 Professor de Vries published a brief account of experiments which he has for several years been carrying on, giving results of the highest value."[10] High, indeed. ". . . I think that I used no extravagant words when, in introducing Mendel's work to the notice of readers of the Royal Horticultural Society's Journal, I ventured to declare that his experiments are worthy to rank with those which laid the foundation of the Atomic laws of chemistry."[10]

Bateson was primed for the rediscovery of Mendel's laws of heredity. The son of the Master of St. John's College at Cambridge, Bateson had distinguished himself as a college student. He was particularly fond of embryology, a field stimulated by Darwin, who pointed out the close re-

1

semblances of the early stages of numerous vertebrate embryos. Embryology was less explored than comparative anatomy or taxonomy, and it gave its practitioner an opportunity to become both a microscopist and an experimentalist. It introduced him to the cell theory directly and forced his attention to the problem of variation at the cellular level. "When the theory of evolution first gained a hearing it was felt that it was of primary importance to know first, whether it was true that forms of life had been evolved from each other, and secondly, if evolved, on what lines had this been effected and what was the ancestry of each. All other problems sank into insignificance in comparison to these. Now the readiest method of answering these questions seemed to be the embryological method . . . the study of embryology superseded all others and elaborate deductions have been made from its results. . . . Embryology has provided us with a magnificent body of facts, but the significance of the facts is still to seek."[24]

At the age of twenty-one, the recipient of a B.A. from St. John's, Bateson came across a circular from Johns Hopkins University announcing the discovery of a marine invertebrate, *Balanoglossus*, at the Chesapeake Bay Zoology Station. Bateson was awarded a fellowship to travel to the United States and was delighted that W. K. Brooks, the director of the station, had given him permission to study *Balanoglossus*. This was an ideal organism for Bateson to study. He was interested in the problem of evolution; and *Balanoglossus*, while not a vertebrate, was nevertheless related to the vertebrates; it possessed characteristics which classified it as a chordate. This enabled Bateson to place *Balanoglossus* in an evolutionary relationship with other primitive chordates.

More significant, however, were Bateson's two summers of association with W. K. Brooks in 1883 and 1884. Brooks, himself an embryologist and student of the world renowned naturalist Agassiz, was an ardent evolutionist who argued strongly that the mechanisms of evolution should be studied through variations. Brooks was not an orthodox Darwinist; he believed that the infinitesimally minor fluctuating variations, which Darwin suggested, might not be the exclusive source for evolution. Bateson was strongly attracted to this view and Brooks's influence remained through the rest of Bateson's life. "For myself I know that it was through Brooks that I first came to realize the problem which for years has been my chief interest and concern. At Cambridge in the eighties morphology held us like a spell. That part of biology was concrete. The discovery of definite, incontrovertible fact is the best kind of scientific work, and morphological research was still bringing in facts in quantity. It scarcely occurred to us that the supply of that particular class of facts was exhaustible, still less that facts of other classes might have a wider significance. In 1883 Brooks was just finishing his book 'Heredity' and naturally his talk used to turn largely on this subject. He used especially to recur to his ideas on the nature and causes of variation. . . . The leading thought was that which he expresses in his book . . . that 'the obscurity and complexity of the phenomena of heredity afford no ground for the belief that the subject is outside the legitimate province of scientific enquiry.' To me the whole province was new. Variation and Heredity with us had stood as axioms. For Brooks they were problems."[19]

During the next ten years Bateson amassed numerous cases of marked differences in varieties and individuals which he designated "discontinuous variations." "He ransacked museums, libraries, and private collections; he attended every sort of 'Show,' mixing freely with gardeners, shepherds, and drovers, learning all they had to teach him. Thus in one way or another, he accumulated a vast store of facts on variation and of practical knowledge of animal and plant life; and to the end of his life, by the same means, he continued ever to add to this store."[24] The Darwinian or continuous variations of a fluctuating sort were still considered the basis of evolution by most biologists. In a forceful statement of this alternative possibility, Bateson published his *Materials for the Study of Variation* in 1894. It was provocative, but neither a financial nor professional success; nevertheless, it led to his election as a Fellow of the Royal Society that same year. Bateson's views met serious competition from a rival school of evolutionists who upheld Darwin's continuous variations as the materials for Natural Selection. This school, founded by Francis Galton and developed by Karl Pearson, called itself the biometrical school. Their tools were mathematical; they developed the statistical analysis of populations. From these studies the biometric school believed that it had clinched the case for continuous variation. Quantitative traits within populations tended to fall into a bell-shaped probability curve. Even here Bateson hoped to find evidence for discontinuous variation. He studied the size of pincher claws in earwigs (small insects commonly found in Europe) and found that a bimodal curve, rather than the expected unimodal curve, was characteristic of a population he sampled.

Nevertheless the mechanism of discontinuous variation was unknown. How was one to approach the species problem experimentally? For in that approach, he felt, the answers would be revealed. He was aware that the success of Darwin's theories came from his careful habits of inquiry; Darwin had a passion for collection. This zeal had permeated the universities and their naturalists. In an appeal to redirect the habits of collectors, Bateson cited the case of the moth *Amphidasys betularia*, hailed by entomologists as a case of evolution in progress. In 1840 a black variety of this light gray speckled moth was noticed in Manchester, and its spread was recorded in other urban areas of Great Britain. This black variety was then designated as a new species, *A. doubledayaria*. To the Darwinist the change was gradual and the moth was surmised to have passed through a series of progressively darker grays. To Bateson the change might have been, and probably was, an instance of discontinuous variation with the dark or melanic form arising suddenly as a "sport" and becoming established as adaptively suited to the sootier urban areas. What concerned

Bateson most, however, was that the problem could have been solved by the collectors themselves if they had adopted the statistical methods of Galton's school and sampled the population randomly (he even suggested using female moths to attract males and thus avoid the personal bias of the collector). But collectors were in the habit of saving the best specimens and destroying the imperfect. This, claimed Bateson, made their collections worthless as contributions to the study of variation and evolution.

Bateson pleaded with naturalist and collector to understand each other's goals. "In order even to choose subjects for his inquiry, still more in order to pursue them, the student of evolution needs the peculiar knowledge and experience—the whole apparatus in fact,—which only the collector possesses. In all this he is too often deficient. It is much if the very names of common objects of natural history are familiar to him, and the world of 'good species' and 'bad species' is unknown. Seldom even can the two classes of men greatly help each other. The collector finds the evolutionist ignorant of what he regards as the rudiments; he only vaguely perceives the other's purposes and is not greatly interested in them. His collection was made with different objects altogether, and though with the best will in the world he puts it at the service of anyone who will work at it, he cannot make it serve a purpose for which it was not designed. The other leaves disappointed. His questions are mostly unanswered, and he is tempted to feel that the methods of the collector are narrow and that he has missed his opportunities. Now both men are right. The future is with the evolutionist, but it is the collector who has made it possible for him to begin his work."[6]

Specifically, Bateson pointed the way he hoped the collector would work. "It is depressing to see how those who are engaged in the business of systematic work often neglect to give the essential particulars as to the variability of the material submitted to them for description. That such a character is 'variable' or 'so variable that no reliance can be placed on it' is often all that we are told when in many cases with little additional trouble the number of specimens exhibiting each variation could have been recorded, thus greatly lightening the task of those who come after. If collectors and systematists would arrange their work in such a way as to bring out and not conceal the objective phenomena of evolution, and if the evolutionist would appreciate that the proper way to study the relation of type and variety is to take up the work at the place where the systematist leaves it, we should have that partnership between the two classes of naturalists for want of which so much effort is wasted and progress is so slow."[6]

The collector and systematist were not the only scientists whose help he solicited. Bateson had a strong conviction that the horticulturist and breeder had much to contribute. Bateson, like Darwin, was aware of the varieties that existed or appeared de novo in animal and plant species of domesticated forms. Perhaps through the study of individual traits this greater problem of variation could be approached. In the summer of 1899, eight months before the first publication of Hugo de Vries's rediscovery of Mendel's findings had been put to press, Bateson addressed an appeal to members of the Royal Horticultural Society. "The recognition of the existence of discontinuity in variation, and of the possibility of complete or integral inheritance when the variety is crossed with the type is, I believe, destined to simplify to us the phenomenon of evolution perhaps beyond anything that we can foresee. At this time we need no more *general* ideas about evolution. We need *particular* knowledge of the evolution of *particular* forms. What we first require is to know what happens when a variety is crossed with its *nearest allies*. If the result is to have a scientific value, it is almost absolutely necessary that the offspring of such a crossing should be examined *statistically*. It must be recorded how many of the offspring resembled each parent and how many showed characters intermediate between those of the parents. If the parents differ in several characters, the offspring must be examined statistically, and marshalled, as it is called, in respect of each of those characters separately."[8]

Bateson himself was now engaged in a variety of crosses which he hoped would illustrate such discontinuities. He was, in fact, on the verge of discovering for himself what Mendel had observed some thirty-five years before. Prophetically, he spoke to his audience: "It is perhaps simplest to follow the beaten track of classification or of comparative anatomy, or to make for the hundredth time collections of the plants and animals belonging to certain orders, or to compete in the production and cultivation of familiar forms of animals or plants. But all these pursuits demand great skill and unflagging attention. Any one of them may well take a man's life. If the work which is now being put into these occupations were devoted to the careful carrying out and recording of experiments of the kind we are contemplating, the result . . . would in some five-and-twenty years make a revolution in our ideas of species, inheritance, variation, and the phenomena which go to make up the science of Natural History. We should, I believe, see a new Natural History created."[8]

A few months after his address to the Horticultural Society, Bateson prepared a second paper extending his theme that virtually nothing was known about the mechanisms of heredity and variation. "We want to know the whole truth of the matter; we want to know the physical basis, the inward and essential nature, the causes, as they are sometimes called, of heredity. We want also to know the laws which the outward and visible phenomena obey. . . ."[9] But Bateson had no such laws to offer; he did have a missionary's zeal to stimulate a search for them and a willingness to confess his total ignorance. "Let us recognize from the outset," he wrote, "that as to the essential nature of these phenomena we still know absolutely nothing. We have no glimmering of an idea as to what constitutes the essential process by which the likeness of the parent is transmitted to the offspring. We can study the process of fertilization and development in the finest detail which the microscope manifests to us, and we may fairly say that we have now a thorough grasp of the visible phenomena; but of the nature of the physical basis of heredity we have no conception at all. No one has yet any suggestion, any working hypothesis, or mental picture that has thus far helped in the slightest degree to penetrate beyond what we see. We do not know what is the essential agent in the transmission of parental characters, not even whether it is a material agent or not. Not only is our ignorance complete, but no one has the remotest idea how to set to work on that part of the problem."[9]

Bateson had reason to expect his audience to sympathize with his view. This would not have been true if he were speaking to a society of botanists or zoologists. Certainly the new school of biometry, led by Galton and Pearson, would have looked upon this statement as unduly pessimistic. They had, after all, proposed a statistical model of heredity which was widely acclaimed throughout Britain as the tool which Darwin might have taken pride in, had he lived to see its application to evolution.

So too, a congress of zoology or botany in continental Europe would have rejected Bateson's pessimism. Hadn't Weismann disproved the older, Lamarckian theory of acquired characteristics? Hadn't Weismann and Roux built a case for "determinants" or other particulate units which could be justified by observations in embryonic development? Even Hugo de Vries had "solved" the weakness of Darwin's theory of pangenesis, combining Weismann's theories with Darwin's particulate units and then limiting them to individual cells. And if these did not suffice, there were additional theories proposed by Spencer and by Weismann's teacher, Nägeli. Each certainly had their adherents and their works were frequently translated. But of these, none was based on experiment. All suffered from being "general ideas about evolution." It was better to confess ignorance; perhaps breeders, using some of the discipline Bateson urged on the naturalists and taxonomists, would have more to offer than their more famous colleagues at the universities.

On May 8, 1900, Bateson boarded the train for his second trip to the meetings of the Horticultural Society. He had his manuscript prepared and he brought along some technical journals to read. One of them contained an account of Mendel's laws which Hugo de Vries had just confirmed. This was Bateson's first encounter with Mendel's experiments. He was gripped by their simplicity, their approach, their implications, and their value for solving the problem of discontinuous variation. He later listed this experience as one of the half dozen most emotional moments of his life. He altered his manuscript to incorporate this astounding new finding, so sure was he of its veracity and significance. Now he could offer more than hope to his audience. "Professor de Vries begins by reference to a remarkable memoir by Gregor Mendel, giving the results of his experiments in crossing varieties of *Pisum sativum*. These experiments of

Mendel's were carried out on a large scale, his account of them is excellent and complete, and the principles which he was able to deduce from them will certainly play a conspicuous part in all future discussions of evolutionary problems."[9]

What were these principles which Mendel found? First, Mendel observed that certain types of characters, ones that could be sharply distinguished—"differentiating characters"—would show an apparent dominance of one character over the other in the progeny of a cross between "differentiated" parents. But the disappearance of one of these characters was not real; it reappeared in the subsequent generation when the progeny were crossed to one another. This meant that the two characters retained their purity and were transmitted as two different "elements." One of these "elements" Mendel called the *dominant* type of character; the other he called *recessive*. "Those characters which are transmitted entire, or almost unchanged in the hybridization, and therefore in themselves constitute the characters of the hybrid, are termed the *dominant*, and those which become latent in the process, *recessive*. The expression 'recessive' has been chosen because the characters thereby designated withdraw or entirely disappear in the hybrids, but nevertheless reappear unchanged in their progeny. . . ."[192] The reappearance of the characters in unaltered form from the hybrid was the most significant feature of this experiment, and de Vries himself recognized this in entitling his paper "On the law of disjunction of hybrids." Mendel also emphasized this as the main feature of his seven years of experimentation with peas: "If two plants which differ constantly in one or several characters be crossed, numerous experiments have demonstrated that the common characters are transmitted unchanged to the hybrids and their progeny. . . ."[192]

Second, Mendel recognized that each such pair of "differentiating characters" acted independently of other pairs of characters that he had used in his experiments. Thus one parent might have all dominant traits, the other might be all recessive. The progeny descending from the hybrids would be mixtures of all combinations of the various dominant or recessive traits. "With *Pisum* it was shown by experiment that the hybrids form eggs and pollen cells of *different* kinds, and that herein lies the reason of the variability of their offspring."[192] With this Bateson could certainly see a mechanism for maintaining discontinuous variation! Mendel recognized that the basis of this type of heredity was a particulate one, and he referred to "elements" in the cells which give rise to his "differentiating characters." "In the formation of these cells all existing elements participate, in an entirely free and equal arrangement, by which it is only the differentiating ones which mutually separate themselves. In this way the production would be rendered possible of as many sorts of egg and pollen cells as there are combinations possible of the formative elements."[192]

All that Bateson hoped for in breeding analysis was present in Mendel's experiments. Individuals, not populations, were used; the traits were discontinuous; and accurate accounts were kept of the numbers of progeny from each cross.

From these accurate accounts Mendel was able to derive some mathematical rules of inheritance. The progeny of hybrids differing in one pair of characters gave a ratio in the F_2 of 3 dominant to 1 recessive—or at least this is the ratio which large numbers of progeny approximated. Similarly, when hybrids differed in two pairs of characters, Mendel found that their progeny approximated a 9 : 3 : 3 : 1 ratio, representing, respectively, the class expressing both dominant traits, the two classes expressing only one dominant trait, and the class which was doubly recessive. An extension of this approach to several characters gave Mendel an algebraic notation for describing the various classes of progeny and types of pollen or eggs that would be produced. For Bateson nothing could be more suitable: it took away the monopoly that biometrics had for biology and it gave to the breeder and to the naturalist a precise, quantitative, experimental procedure for the analysis of heredity and variation.

CHAPTER 2

The Fight to Legitimize Genetics

Shortly after Mendel's paper was popularized by the appearance of the articles of de Vries, Tschermak, and Correns, the biometric school launched an attack against its universality. At Cambridge, W. F. R. Weldon was a strong convert to Galton's theory of ancestral inheritance. Ironically, Weldon was also the embryologist who had stimulated Bateson's interest in *Balanoglossus*; it was Weldon who had arranged the fellowship which enabled him to go to Brooks's laboratory on Chesapeake Bay. But as Bateson amassed evidence for discontinuous variation, Weldon grew more critical about these unorthodox views of evolution. Weldon accepted the Darwinian view that change was gradual; and Galton's insight that mathematics could describe this variation in populations was sufficient to convince Weldon that Bateson was following an artificial and improbable mechanism for the origin of variations.

In the hopes of interesting biologists in their fledgling science of biometrics, Pearson and Galton founded their own journal, *Biometrika*, in 1901. In their foreword, asserting the purposes and scope of their venture, the editors reaffirmed their faith in population analysis as the source of the laws of heredity. The journal is dedicated to Darwin, and a marble statue of him appears saint-like in a photograph cropped to resemble a gothic church window. Weldon, disturbed by a rival theory of heredity which Bateson had claimed would supersede Galton's, had a difficult task. Was Mendelism or Galtonism the more valid approach to the study of heredity? Assured of his verdict after critically reading the Mendelian literature and

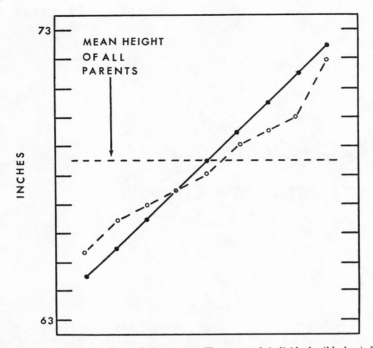

Figure 1. *Galton's Law of Regression.* The parental individuals (black circles) give rise to offspring (open circles) of less extreme height than the parents. Galton applied this law to other quantitative traits, including intelligence, and advocated a eugenic program to escape the trend to mediocrity.

even using Bateson's technique of going to the breeders, Weldon, through the pages of *Biometrika*, began his attack. "Now it is well known to all breeders, and it is clearly shown in a number of cases by Galton and Pearson, that the condition of an animal does not as a rule depend upon the condition of any one pair of ancestors alone, but in varying degrees upon the condition of all its ancestors in every past generation, the condition in each of the half dozen nearest generations having a quite sensible effect."[318]

In contrast to this prevailing view, Mendel's theory of alternative inheritance attributes to the parents alone the source of their progeny's heredity. But "if Mendel's statements were universally valid, even among peas, the character of the seeds in the numerous hybrid races now existing should fall into one or another of a few definite categories, which should not be connected by intermediate forms. . . . I have carefully examined the seed characters of some twenty named varieties, and the present condition of many I have studied seems to me quite incompatible with the general validity of Mendel's statements."[318] Weldon cited his own observations and those of many breeders who had examined degrees of wrinkling in seed coats; they found some commercial varieties whose wrinkling formed a continuous curve from barely perceptible dents to sharply corrugated wrinkles. Many of the round peas themselves showed occasional dents although these were from strains which never segregated wrinkled peas. Similarly, yellow strains showed degrees of color from chlorotic tints to full green, often with piebald patches of green interspersed with yellow on the surface of the cotyledons. Was this the vaunted separability of characters which the Mendelists hoped to use? "Enough has been said to show the grave discrepancies between the evidence afforded by Mendel's own experiments, and that obtained by other observers, equally competent and trustworthy. . . . The evidence brought together rests upon the statements of men whose knowledge and skill are beyond question, and the only conclusion which can, I think, be fairly drawn from it is that dominance of any of the characters mentioned is not an invariable attribute of the character, but that a cross between pairs of parents . . . may in different cases lead to widely different results. . . ."[318]

This was critical evidence; it demanded experimental work of comparable quality to restore the universality of Mendel's laws to the problem of heredity. If Weldon had stopped here he might have won the battle and delayed the inevitable victory of Mendelism for more years than it took. But Weldon instead chose to add wrath to criticism and to bolster his own faith in biometrics as the exclusive approach: "the fundamental mistake which vitiates all work based upon Mendel's method is the neglect of ancestry, and the attempt to regard the whole effect upon offspring, produced by a particular parent, as due to the existence in the parent of particular structural characters; while the contradictory results obtained by those who have observed the offspring of parents apparently identical in certain characters show clearly enough that not only the parents themselves but their race, that is their ancestry, must be taken into account before the results of pairing them can be predicted."[318]

Bateson was never one to decline a debate. However, Weldon was his benefactor and his senior. Bateson was now forty years old; he had not yet received a teaching position or permanent research position with Cambridge University. He was still living on an insecure yearly stipend which he had to supplement by applications for occasional small grants. He was, in effect, a postgraduate on fellowship for almost twenty years. If he had chosen to follow the path of *Balanoglossus* he would probably have had his tenure and a Readership or Professorship. His militant attitude against Darwin's interpretation of variation made his superiors hesitate; it antagonized those who would have been his friends. Now Bateson had to choose his weapons: a reply in *Biometrika*? This would have been reasonable but Bateson wanted a larger audience; he did not want to reply to Weldon, he wanted Mendelism to become the outlook for twentieth century Darwinism! Bateson chose to write a book—*A Defence of Mendel's Principles of*

Heredity. Here, free of an editor's censoring pencil, he could form his case for Mendel. In doing this he would return sarcasm for wrath. "Therefore on the announcement of that discovery which once and for all ratifies and consolidates the conception of discontinuous variation, and goes far to define that of alternative inheritance, giving a finite body to what before was vague and tentative, it is small wonder if Professor Weldon is disposed to criticism rather than cordiality."[12] Far from conceding a weakness in Mendelism, Bateson underlined its potential. "On the rediscovery and confirmation of Mendel's Law by de Vries, Correns, and Tschermak two years ago, it became clear to many naturalists, as it certainly is to me, that we had found a principle which is destined to play a part in the study of Evolution comparable only with the achievement of Darwin—that after the weary halt of forty years we have at last begun to march."[12]

Polemics would not convince the critical reader that Mendelism was worth studying. Bateson first had to state his position. If Mendelism was universal, what is the significance of Galtonism? "Whether the Mendelian principle can be extended so as to include some apparently Galtonian cases is another question, respecting which we have as yet no facts to guide us, but we have certainly no warrant for declaring such an extension to be impossible."[12] This was not a concession for a coexistence of two theories of heredity. "With the distinctions between the original Law of Ancestral Heredity, the modified form of the same law, and the law of Reversion, important as all these considerations are, we are not at present concerned. For the Mendelian principle of heredity asserts a proposition absolutely at variance with all the laws of ancestral heredity, however formulated. In those cases to which it applies strictly, this principle declares that the cross-breeding of parents need not diminish the purity of their germ-cells or consequently the purity of their offspring."[12]

But if there is no compromise on the rival views, how does one explain the continuity observed in the commercial peas for seed color and degree of wrinkling? To this Bateson confesses an earlier oversimplification. "It is difficult to blame those who on first acquaintance concluded Mendel's principle can have no strict application save to alternative inheritance. Whatever blame there is in this I share with Professor Weldon and those whom he follows. Mendel's own cases were almost all alternative; also the fact of dominance is very dazzling at first. But that was two years ago, and when one begins to see clearly again, it does not look so certain that the real essence of Mendel's discovery, the purity of the germ cells in respect of certain characters, may not apply also to some phenomena of blended inheritance."[12]

The book is a woven collection of his address to the Horticultural Society, of Mendel's principles (including a translation of Mendel's original paper), and most important, an edited account of the detailed experiments carried out in his own gardens and reported in a 160-page report to the Evolution Committee of the Royal Society. The total view is not a report of what Mendel had done but of what could be done and what was then in progress to extend Mendelism to both the plant and animal kingdoms. "I trust that what I have written has convinced the reader that we are, as we said in opening, at last beginning to move. Professor Weldon declares that he has 'no wish to belittle the importance of Mendel's achievement'; he desires 'simply to call attention to a series of facts which seem to him to suggest fruitful lines of inquiry.' In this purpose I venture to assist him, for I am disposed to think that unaided he is—to borrow Horace Walpole's phrase—about as likely to light a fire with a wet dish-clout as to kindle interest in Mendel's discoveries by his tempered appreciation. If I have helped a little in this cause my time has not been wasted."[12]

Bateson's Contribution to Mendelism

Bateson recognized that the original form of Mendel's paper was written in the conceptual view of the mid-nineteenth century. Between this period of time and his own, the cell theory had been put on a sound and irrefutable foundation; physiology, biochemistry, histology, and cytology had come into being, and the new disciplines made the organism a more complex system. Biologists coped with the new discoveries by innovating their own terms, not as jargon but as a simplification of tediously descriptive phrasing. With this historical precedent as an acceptable procedure, Bateson tried to simplify Mendelism. In a short paragraph in his first report to the Evolution Committee in 1902, Bateson introduced four terms in rapid order: "By crossing two forms exhibiting antagonistic characters, cross-breds were produced. The generative cells of these cross-breds were shown to be of two kinds, each being pure in respect of one of the parental characters. The purity of the germ-cells, and their inability to transmit both of the antagonistic characters, is the central fact proved by Mendel's work. We thus reach the conception of unit-characters existing in antagonistic pairs. Such characters we propose to call *allelomorphs*, and the zygote formed by the union of a pair of opposite allelomorphic gametes we shall call a *heterozygote*. Similarly, the zygote formed by the union of gametes having similar allelomorphs, may be spoken of as a *homozygote*."[26]

Almost thirty years later allelomorph was abbreviated to *allele*. The terms homozygote and heterozygote are still in use today in their original sense. But the one term *not underlined* by Bateson, taken almost as an obvious abstraction of the facts of inheritance, was a conceptual disaster. The *unit-character* combined the "differentiating character" used by Mendel with the "formative element" which he assumed to represent it in the germ cell. Although Mendel was not consistent in keeping these two concepts separate, he did not merge the two, as Bateson did. In this hyphenated form, the unit-character conveyed the impression that the character, and not some "formative element" or unit, was inherited.

Bateson had an awareness that the character should not be confused with the basis for its transmission. It was just such an identification by Weldon which led him to think that the seed color of peas was not alternative but was usually blending in inheritance. Bateson was not confused by this identification, but for the next fifteen years this fallacy would create a crisis among students of heredity.

Bateson also had difficulty putting down his results in comprehensible form. To write out the mating procedure and give the results of experiments in long, complex sentences was tedious for the author and more stifling to his readers. To avoid this Bateson suggested that "it is absolutely necessary that in work of this description some uniform notation should be adopted. Great confusion is created by the use of merely descriptive terms, such as 'hybrid generation,' 'second generation of hybrids,' etc., and it is clear that even to the understanding of the comparatively simple cases with which Mendel dealt, the want of some such system has led to difficulty. In the present paper we have followed the usual modes of expression, but in future we propose to use a system of notation modelled on that used by Galton in 'Hereditary Genius.' We suggest as a convenient designation for the parental generation the letter P. In crossing, the P generation are the pure forms. The offspring of the first cross are the first filial generation F. Subsequent filial generations may be designated by F_2, F_3, etc. Similarly, starting from any subject individual, P_2 is the grandparental, P_3 the great-grandparental generation, and so on."[26] Later, Bateson and his colleagues used P_1 for the immediate parental generation and F_1 for their progeny, the first filial generation.

More than nomenclature was needed. Bateson recognized from the work that he and his students had been carrying out since 1897 that unusual cases could be explained by Mendelism if one granted that "the fact that in the cross-breds one character, in appearance, dominated to the exclusion of the other is not of the essence of Mendel's discovery."[26] One such exception occurred in his poultry crosses. A heterozygous form (blue Andalusian) consistently differed from either of the homozygous parents. "There is, therefore, a strong probability that the Andalusian is a heterozygote, though, doubtless, of a complex nature. Its gametes do not fully correspond to it, and its colour must be produced by a combination of dissimilar allelomorphs."[26] Formerly breeders had called this "instability" and now Bateson attributed the phenomenon to a heterozygous appearance differing from that of either homozygote.

Previously he alluded to blending inheritance as a special case of the independent inheritance of several unit-characters. He tried to extend this possibility to a specific case. "It must be recognized that in, for example, the stature of a civilized race of man, a typically continuous character, there must certainly be on any hypothesis more than one pair of possible allelomorphs. There may be many such pairs, but we have no certainty that the number of such pairs and consequently of the different kinds of gametes are altogether *unlimited* even in regard to stature. If there were even so few as, say, four or five pairs of possible allelomorphs, the various homo- and heterozygous combinations might, on seriation, give so near an approach to a continuous curve, that the purity of the elements would be unsuspected, and their detection practically impossible. Especially would this be the case in a character like stature, which is undoubtedly very sensitive to environmental accidents."[26] This view, foreshadowing the theory of "multiple factors" by several years, would make Galton's observed phenomena, but not his interpretations, acceptably Mendelian.

Among other implications which Bateson foresaw for Mendelism were the inheritance of lethal factors ". . . a recessive allelomorph may even persist as a gamete without the corresponding homozygote having ever reached maturity in the history of the species."[26] As an example of such a subvital trait, Bateson cited Garrod's observation that children with alkaptonuria and albinism are more frequently found in marriages among first cousins than in the general unrelated population. This coincidence, Bateson inferred, was consistent with the idea that the trait was a recessive and that in other disorders (inherited in a similar way) the severity of the homozygous defect would prevent its victims from transmitting the disease, but its absence of effect on the parents meant that it could be perpetuated through heterozygotes. In addition, Bateson attempted to explain permanent hybrids (which would be detected by outcrossings to other varieties) by a system similar to "balanced lethals" which would be suggested formally about fifteen years later. "The existence of forms which are exclusively heterozygous leads to the contemplation of another possibility. In the heterozygotes we have spoken of, both sexes of course bear gametes transmitting each allelomorph. If, however, one allelomorph were alone produced by the male and the other by the female we should have a species consisting of *only* of heterozygotes."[26]

The appearance of defective "rogues" among selected strains of crop plants he attributed to occasional homozygous recessive forms similar to the cases of alkaptonuria in man. So-called "reversions" to an ancestral type he attributed to complex characters which superficially looked alike but whose hereditary basis resided in different pairs of allelomorphs. Crosses between such strains resulted in the coming together of dominant unit-factors that had been separately maintained in the two varieties. This gave a new significance to purity, which "then acquires a new and more precise meaning. An organism resulting from an original cross is not necessarily pure when it has been raised by selection from parents similar in appearance for an indefinite number of generations. It is only pure when it is compounded of gametes bearing identical allelomorphs, and such parts may occur in any individual raised from cross-bred organisms."[26]

All these thoughts and the new discoveries which accumulated in the next three years made Bateson a figure to be respected, even if disliked. His income was still low and uncertain. He was not honored with an appointment to the teaching staff. The new Mendelism was not being taught at the university. But in 1905, Cambridge University received a bequest from the Quick estate which was to be used to establish a chair in a field of Zoology which showed future promise. Bateson was asked what title he would choose if the Quick Fund were used for him. In reply, Bateson wrote to his colleague, Professor A. Sedgwick, "If the Quick Fund were used for the foundation of a professorship relating to Heredity and Variation the best title would, I think, be 'The Quick Professorship of the study of Heredity.' No single word in common use quite gives this meaning. Such a word is badly wanted, and if it were desirable to coin one, 'GENETICS' might do. Either expression clearly includes variation and the cognate phenomena."[14] Early in the next year Bateson was disappointed to learn that the Quick Fund would be used to establish a Professorship in *Proto-zoology*. Despite his disappointment, Bateson did not discard his new term; he informed an International Congress of Botany that "a new and well developed branch of Physiology has been created. To this study we may give the title *Genetics*."[17]

CHAPTER 3

The Need for Hereditary Units

The recognition by Mendel that germinal "elements" transmitted the "differentiating characters" of his pea plants was one of the first attempts to represent an organism as the composite of particulate hereditary determinants. Mendel came to this abstraction through the experimental discovery of segregation. Before the rediscovery of Mendelism, several theories of heredity were proposed which had, as their major premise, particulate units serving as hereditary determiners. None of these theories came directly from experimental findings; some owed their origin to Darwin's speculations; others developed from cytological studies of cell nuclei and the chromosomes. Whatever their origin, none had predictive value for laboratory tests. They served, nevertheless, as a philosophical climate which made the rediscovery of Mendelism not only acceptable, but comprehensible. This change in outlook was one of the differences that made Mendelism obscure in 1865 and renowned in 1900.

The need for units was axiomatic among students of development and cell biology. This prevailing view, expressed by E. G. Conklin in an address to the American Philosophical Society in 1898, asserted that "the organization of the cell . . . does not stop with what the microscope reveals, but must be supposed to extend to the smallest ultimate particles of living matter which manifest specific functions. These are the vital units so generally postulated, the 'smallest parts' of living matter, as they were called by Brücke, who first demonstrated that they must exist; the 'physiological units' of Spencer, the 'gemmules' of Darwin, the 'micella groups' of Nägeli, the 'pangenes' of de Vries, the 'plasomes' of Weisner, the 'idioblasts' of Hertwig, the 'biophores' of Weismann. Such ultimate units have been found absolutely necessary to explain those most fundamental of all vital phenomena, *assimilation* and *growth*, while many other phenomena, especially particulate inheritance, the independent variability of parts, and the hereditary transmission of *latent* and *patent* characters, can at present only be explained by referring them to ultra-microscopic units of structure."[86]

Darwin had argued that Natural Selection was dependent on variation. He believed that the variation in a breeding population was of a fluctuating nature, with selection choosing the most suitable forms among the spectrum of variation. But what was the mechanism for this variation? It had to be inheritable; otherwise selection would be meaningless for the progress of evolution. Darwin attempted experiments with pigeons and kept himself informed of the results of breeding experiments performed with various domesticated plants and animals. Unfortunately, Darwin studied characters that were complex rather than sharply alternative. His characters tended to blend and form a quantitative graded series which re-enforced his conviction that the variations in nature were of almost imperceptible fluctuating degrees. As a working theory, Darwin suggested microscopic or ultramicroscopic particles, the gemmules, which existed in all cells. These responded to the physiological conditions of the tissues in which they subsisted and thus could be varied by the external environment. Their tissue source, their numbers, the time at which they were released from their cells into the circulatory system to find their way to the gonad, and the stresses imposed by the environment resulted in the uniqueness of the individual gamete, which was a composite of gemmules; they also generated the differences in traits, the character variations, on which Natural Selection could work.

The appeal of this theory of Pangenesis, as Darwin called it, was most strongly felt by Hugo de Vries. Prior to his rediscovery of Mendelism he proposed a theory in 1889, based on Darwin's pangenesis but modified to account for the experimental evidence against the inheritance of acquired characteristics. This new theory he designated "Intracellular Pangenesis." "The visible phenomena of heredity are . . . the expressions of the characters of minutest invisible particles, concealed in that living matter. And we must, indeed, in order to be able to account for all the phenomena, assume special particles for every hereditary character. I designate these units, pangens."[110] De Vries did not want to fall into a trap that his pangens were a disguised form of the discredited idea of preformation; "these pangens do not each represent a morphological member of the organism, a cell, or a part of a cell, but each a special hereditary character. . . . The pangens are not chemical molecules, but morphological structures, each built of numerous molecules. . . . We must simply look for the chief life-attributes in them, without being able to explain them. We must therefore assume that they assimilate and take nourishment and thereby grow, and then multiply by division; two new pangens, like the original one, usually originate at each cleavage. Deviations from this rule form a starting point for the origin of variations and species." Like Weismann and Roux, de Vries assumed a developmental significance for his units. "The differentiation of the organs must be due to the fact that individual pangens or groups of them develop more vigorously than others. The more a certain group predominates, the more pronounced becomes the character of the respective cell."[112]

The variation required in Darwin's theory was attributed to gemmules subject to modification by the environment. To avoid this Lamarckian interpretation, de Vries denied that pangens could be altered by the environment and he eliminated as unnecessary the "carrying mechanism" of the particles from the somatic cells to the gonads. If the germinal pangens were separated from the somatic pangens because all pangens remained *intracellular*, then the units themselves became the most significant feature of heredity. "According to pangenesis, there may be two kinds of variability. These are differentiated in the following manner by Darwin. In the first place the pangens present may vary in their relative number, some may increase, others may decrease or disappear almost entirely; some that have long been inactive may resume activity and finally the grouping of the individual pangens may possibly change. All of these processes will amply explain a strongly fluctuating variability. . . . In a word, an altered numerical relation of the pangens already present, and the formation of new kinds of pangens must form the two main factors of variability."[112] It is interesting that this last sentence contained the idea that provoked de Vries into a search for suitable material for the study of discontinuous variations, or *mutations* as he called them, and this study in turn led to his rediscovery of Mendelism.

The nature of these pangens was unknown, but de Vries agreed with "the majority of investigators (who) assume that the material bearers of hereditary characters are units, each of which is built up of numerous chemical molecules, and is altogether a structure of another order than the latter."[112] Furthermore, such complex molecular pangens were inferred to be "latent" in the nucleus and "active" in the cytoplasm. A body of evidence was gathering for designating the nuclei of cells as "the reservoirs of hereditary characters"; and, specifically, "most investigators regard the chromatic thread as the morphological place where the material bearers of the hereditary qualities are stored. This thread would, therefore, consist of pangens united into smaller or larger groups, and it shows, in its thickest portions, a distinct structure of special particles strung together. We can entirely agree with the opinion of Roux, where he sees, in the longitudinal splitting of the nuclear skein, the visible part of the segregation of maternal factors into the two halves destined for the two daughter cells."[112]

Bateson, while aware of the developmental, cytological, and physiological possibilities for a particulate theory of heredity, did not try to press the unit-character in the same direction as Roux, Weismann, or de Vries. It was sufficient for him to make a general prediction encompassing this view, but he was too conservative to jeopardize the unit-character with more than he could visualize in experimental form. "It is reasonable to infer that a science of Stoechiometry will now be created for living things, a science which shall provide an analysis, and an exact determination of their constituents. The units with which that science must deal, we may speak of, for the present, as character-units, the sensible manifestations or physiological units of as yet unknown nature."[26]

What Bateson did see, however, was the assimilation of Galton's cases of blending inheritance into a Mendelian scheme. The experimental evidence for this first appeared in 1903 with Johannsen's analysis of quantitative characters in beans. Johannsen claimed that selection in homozygous forms (inbred "pure lines") was ineffectual, and that hybrid variability could be represented as a series of such "pure lines," each of which could be isolated by inbreeding. The "pure line doctrine" gave Bateson a new insight into his unit-characters. "We suggested 'that Galton's law may be a representation of particular groups of cases which are, in fact, Mendelian,' in the sense that the gametes are pure. The analysis carried out by Yule points to a similar conclusion, if the phenomena of dominance and special consequences of heterozygosis are neglected. Pearson's conclusion that various phenomena of inheritance studied by him are incompatible with Mendelian expectation is open to the objection that many of his characters are obviously liable to such great disturbances from the interference of conditions that the operations of heredity alone must be largely obliterated. . . . The experiments of Johannsen have made an important contribution to this part of the inquiry. Taking self-fertilized beans (Phaseolus) he found indications that the number of 'pure lines' in respect of seed-weight was very considerable. Hence we may suppose with some confidence that segregation, in respect of this character, deals with units which, though small, have a sensible size."[27]

The concept of the unit-character, while useful to the experimentalist, was inadequate to represent the units Johannsen had in mind. The characters he studied were complex and many physiological units were involved in the formation of the total character. For this reason Johannsen sought to replace the unit-character and the earlier particulate units of Spencer, Darwin, Nägeli, Weismann, Roux, and de Vries. The closest unit to what he had in mind was the pangen of Darwin and de Vries. To avoid the historical connotations associated with this term, Johannsen offered an abbreviation: "Therefore it appears simplest to isolate the last syllable, 'gene,' which alone is of interest to us, from Darwin's well known word, and thereby replace the less-desirable, ambiguous word 'determiner.' Consequently, we will simply speak of 'the gene' and 'the genes' instead of 'the pangen' and 'the pangens.' The word 'gene' is completely free from any hypotheses; it expresses only the evident fact that, in any case, many characteristics of the organism are specified in the gametes by means of special conditions, foundations, and determiners which are present in

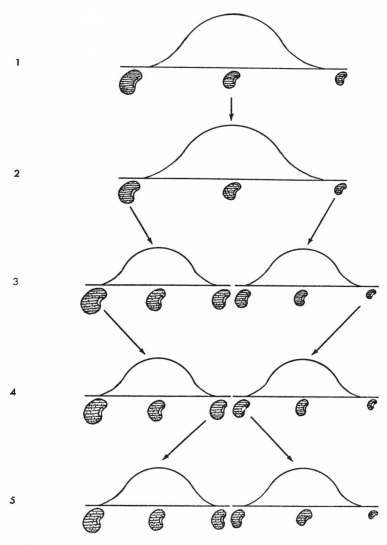

Figure 2. *Johannsen's Analysis of Phaseolus.* In (1) the unselected median bean repeats the normal distribution (2). If selection and inbreeding of large or small beans is initiated (2), the curves establish pure lines (3). Selection of largest beans in the "large" pure line (or smallest beans in the "small" pure line) does not alter the curves (4). This pure line remains unchanged even if the smallest bean in the "large" pure line (or the largest bean in the "small" pure line) is used to generate new attempts at selection within pure lines (5). Johannsen called hereditary constitution of the pure line a *genotype,* and the bean size itself he called the *phenotype.*

unique, separate, and thereby independent ways—in short, precisely what we wish to call genes."[175]

Johannsen's gene was undefined. If to some geneticists he gave a concept that appeared to lack a material reality, for others he freed the concept from any one theory of action, specificity, or composition. It gave the gene concept an opportunity to evolve and to take on, or discard, definition.

CHAPTER 4

The Unit-Character Fallacy

While Bateson fought his critics at home with the hope of establishing Mendelism as the foremost tool for the analysis of heredity, a new storm of criticism began across the Atlantic Ocean. At Harvard University, William E. Castle was engaged in a series of experiments with small mammals to investigate the nature of sexuality. The rediscovery of Mendel's laws and the forceful presentation of Mendel's defense by Bateson encouraged Castle to apply Mendelian studies to these mammals. "It remained, however, for Bateson . . . to point out the full importance and the wide applicability of the law. This he has done in two recent publications with an enthusiasm which can hardly fail to prove contagious. There is little danger, I think, of Mendel's discovery being again forgotten."[63] Castle, following Bateson's example, was presenting Mendelism to the American scientific audience.

In his 1903 address to the American Academy of Arts and Sciences, Castle digested the main findings of Mendel, his rediscoverers, and Bateson. Mendel's main contribution, ". . . this perfectly simple principle, is known as the law of 'segregation' or the law of the 'purity of the germ cell.' "[63] Castle had demonstrated that albinism behaved as a simple recessive in mice and probably was the same in man. Later that year Castle encountered some difficulties in his own work and read similar observations by Darbishire which made him doubt the "perfectly simple" principle of segregation. "Darbishire is unable to regard albinism as beyond qualification recessive, in the Mendelian sense, because white does not entirely disappear from the bodies of the offspring in his first cross."[79] Indeed, Castle's own work showed that a "composite character may undergo resolution into its elements."[79] This might also be true for Darbishire's observa-

23

tions and it would follow ". . . that not all albinos breed alike when crossed
with the same pigmented stock, a conclusion which our own experiments
fully substantiate. This indicates that the gametes formed by different albino stocks are not all alike, and raises the question whether all are equally
pure as regards the pigment forming character."[79] Nevertheless, the essential feature of Mendelism was not to be repudiated. "The Mendelian
doctrine of gametic purity is fully substantiated by experiments in breeding mice, guinea pigs, and rabbits, but with the important qualification
(that) crossbreeding is able to bring into activity latent characters or latent
elements of a complex character."[79] The nature of these latent elements
did not immediately suggest itself to Castle, but in 1905 a more precise
model was offered.

The attempt to justify Mendelian segregation as a partially accurate
process appealed to Castle because it preserved the Darwinian fluctuations in characters as the major basis for selection. By retaining the unit-
character and assuming that the fluctuations occurred in it, he could bring
character variation into a Mendelian framework without invoking additional, unproved, unit-characters. "It would appear that in alternative inheritance characters behave as units, and, more than that, as wholly independent units, so that to forecast the outcome of matings is merely a
matter of mathematics. While this is in a measure true, it is, fortunately
or unfortunately, not the whole truth. In alternative inheritance, characters
do behave as units independent of one another, but the union of dominant
character with recessive in a cross-bred animal is not so simple a process as
putting together two pieces of glass, nor is their segregation at the formation
of gametes so complete in many cases as the separation of the two glass
plates. The union of maternal and paternal substance in the germ-cells of
the cross-bred animal is evidently a fairly intimate one, and the segregation
which they undergo when the sexual elements are formed is more like
cutting apart two kinds of differently colored wax fused in adjacent layers
of a common lump. Work carefully as we will, traces of one layer are almost
certainly to be included in the other, so that while the two strata retain
their identity, each is slightly modified by their previous union in a common lump."[64]

As the analogy implies, the alleles must come together and "contaminate" each other. Citing his own work, Castle argued that ". . . cross-
breeding albino guinea-pigs with blacks increases the amount of black pigmentation formed at the extremities by the albinos, and induces a slight
pigmentation of the coat generally. . . . How far the contamination of the
albinos can be carried, I am unable as yet to say."[65]

The "pure line" doctrine of Johannsen, supported by Bateson's conviction that blending could be interpreted along Mendelian lines by invoking more unit-characters, was scarcely affected by Castle's alternative
view of the impurity of unit-characters. Johannsen was continuing the work
he had reported in 1903, and by 1906 he could assert more confidently

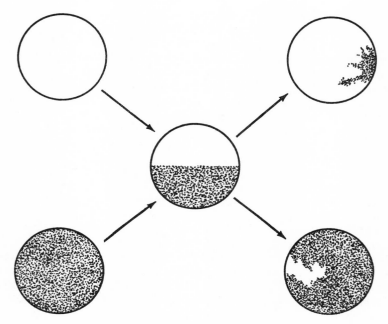

Figure 3. Castle's Contamination Theory. The two pure factors brought into the heterozygous condition emerge relatively pure after meiosis (segregation). A slight contamination of one allele by the other results in the loss of purity of both factors. This is manifested by variability of the progeny, as in the case of hooded rats or spotted rabbits.

that ". . . the study of heredity as to characters, which by inspection can only be estimated as differences in intensity of the same quality, and which blend in hybridisation, requires special methods. The hybrids with such characters have not yet been examined in a satisfactory manner. In my experiments with 'pure lines' I particularly tried to isolate quantitatively different types from the population in question, and in that way I—as the first, I believe—found out that the Galtonian law of filial regression, declaring that fluctuations are to a considerable degree hereditary, is quite wrong and only depends on the presence of several different types in the populations. In a population containing only one single type the selection of fluctuations has no action at all!"[174] It was not to Castle, but to the biometric school that Johannsen addressed his criticisms. Ironically, his main argument would be used by Castle to deny the value of Johannsen's work: "at all events I must again say emphatically that results as to which the analysis has not been fully performed, or cannot be effected, must never be used as a basis for fundamental biological theories. We have always to elucidate the unanalysed from the analysed facts; the converse proceeding is wrong."[174]

Johannsen did not use the concept of the unit-character as the basis of his pure-line population. It would still be another three years before the "gene" would be delivered from the "pangen." To discuss his work, the alternative term "factor" was introduced. Castle rejected the introduction of this term; it was an evasion of the obvious observation that the characters which he studied truly varied. "In several recent papers I have pointed out the fact that the theoretical "purity of gametes" of Mendelian inheritance does not exist. No more does the *purity of factors* exist. We cannot avoid the idea of impurity of the gametes by introducing the conception of 'factors,' for the factors are as certainly impure as the gametes."[65]

Only four years had elapsed since the introduction of the unit-character into the geneticist's vocabulary. It had now produced a philosophical impasse. Why? Castle interpreted the unit-character as a *single* concept. Bateson had tied together both the transmission and the depiction of a character with a hyphen. When unit-character was changed to "unit-factor" or to "factor" alone, Castle no longer dissociated the transmitting agent from its effect on a character. For the next thirteen years Castle would labor with a *unitary* view of the unit-character and its several synonyms. His experiments would be designed, not to *isolate* factors affecting variation, but to *obtain* variation in a character. This was not just a mere blind spot in Castle's thinking. Castle adopted the unitary hypothesis as the better interpretation because it involved fewer assumptions. "To sum the matter up, it is certain that unit-characters exist, but it is equally certain that the units are capable of modification; gametic segregation certainly occurs in some cases (Mendelian inheritance), it does not occur in others (blending inheritance); factors of characters certainly exist, when characters are demonstrably complex and result from the coexistence of two or more simpler ones, as, for example, a purple pigmentation due to coexistence of red and blue chloroplastids in plants. But let us in no case introduce more factors into our hypotheses than can be shown actually to exist."[65]

Bateson was not impressed with Castle's views on the instability of the unit-character. "Castle adheres to the view that pied forms indicate an imperfection of segregation. The propriety of this view becomes less clear in the light of modern evidence proving the existence of determiners for pattern."[17]

Throughout this strong attack on the constancy of unit-characters, Castle avoided introducing a direct or indirect effect of the environment on the instability of unit-characters. Unit-characters were not to be confused with Darwinian gemmules! But in 1907 and 1908 two reports, by Magnus and by Guthrie, claimed that Lamarckian effects did occur when ova of one colored variety of hen or rabbit were transplanted into partially ovarectomized females of another colored variety. Castle, with his student Phillips, was not convinced that these experiments were properly carried out. The possibility of ovarian regeneration was overlooked in both cases. But, "the theoretical importance of this point led us about a year ago to plan experiments which should not be open to the objection which we have stated.

We therefore undertook the transfer of ovarian tissue from a Mendelian dominant to a Mendelian recessive individual. For if, in such a case, germ cells were liberated which bore the dominant character, we should know that they could have come only from the introduced tissue, since recessive individuals are themselves incapable of liberating dominant germ cells. . . . The ovaries were removed from an albino guinea-pig about five months old and in their stead were introduced the ovaries of a black guinea-pig about one month old. The albino upon which the operation had been performed was then placed with an albino male guinea-pig, and six months later bore two black-pigmented young."[82] It was well-established that "albinos when mated with each other only produce albino young. Accordingly there seems no room for doubt that in the case described the black-pigmented young derived their color, not from the albino which bore them, but from the month-old black animal which furnished the undeveloped ovaries for transplantation into the albino. . . . Our results are at variance with those of Guthrie and Magnus. We can detect no modification."[82] Contamination of unit-characters, therefore, had to come from the commingling of gametes and not from the influences of the somatic tissues.

Between 1908 and 1910 two important events occurred. Johannsen published his *Elemente der Exakten Erblichkeitslehre* presenting his genetic researches on the theory of pure lines. The character of an organism he designated as the *phenotype*; the factorial basis for this character he termed the *genotype*; and the factors themselves were now to be called the *genes*. Johannsen, however, had not yet established that several genes or factors were demonstrably present in any single genotype of a pure line. That such a factorial basis could be demonstrated was shown independently by Nilsson-Ehle in Sweden using wheat and by E. M. East in the United States using maize. The "multiple factor" hypothesis which they proposed attributed the inheritance of quantitative characters to two or more *demonstrable* pairs of factors. East asserted that "in my own work there is sufficient proof to show that in certain cases the endosperm of maize contains two indistinguishable, independent yellow colors, although in most yellow races only one color is present. There is also some evidence that there are three and possibly four independent red colors in the pericarp, and two colors in the aleurone cells."[118] These results could be interpreted as a factorial analysis of blending inheritance. "In certain cases it would appear that we may have several allelomorphic pairs each of which is inherited independently of the others, and each of which is separately capable of forming the same character. When present in different numbers in different individuals, these units simply form quantitative differences. . . . When a smaller population is considered, it will appear to be a blend of the two parents with a fluctuating variability on each side of its mode."[118]

The revelations of *proved* multiple factor inheritance forced Castle to reappraise his position. "Size variation is apparently continuous and its inheritance blending, color variation is discontinuous and its inheritance

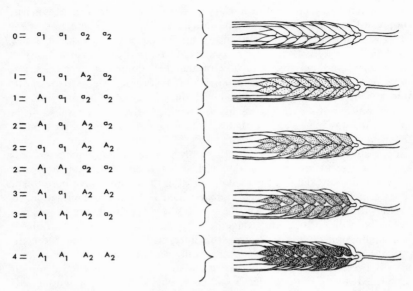

$$0 = a_1 \ a_1 \ a_2 \ a_2$$

$$1 = a_1 \ a_1 \ A_2 \ a_2$$
$$1 = A_1 \ a_1 \ a_2 \ a_2$$

$$2 = A_1 \ a_1 \ A_2 \ a_2$$
$$2 = a_1 \ a_1 \ A_2 \ A_2$$
$$2 = A_1 \ A_1 \ a_2 \ a_2$$

$$3 = A_1 \ a_1 \ A_2 \ A_2$$
$$3 = A_1 \ A_1 \ A_2 \ a_2$$

$$4 = A_1 \ A_1 \ A_2 \ A_2$$

Figure 4. Multiple Factor Inheritance. The color character in the wheat is determined by the number of color factors (A). Colorless wheat has no dominant factors. The dark red wheat contains four such factors. Nilsson-Ehle proposed a multiple factor ("polygenic") inheritance for quantitative traits. This was independently proposed by East.

Mendelian. Notwithstanding the seemingly radical difference between these two types of inheritance, it is possible that they may, after all, prove to have a common basis. Blending inheritance may possibly be only a complex sort of Mendelian inheritance, in which many independent factors are simultaneously concerned. The question is one of much theoretical interest. Its solution awaits further investigation."[66] Gradually, however, Castle became skeptical of the significance of multiple factors as the basis of fluctuations in populations: "It is impossible to deny the soundness of the reasoning of Johannsen and Jennings. It is perfectly clear that the effects of selection *should* be more immediate and much greater in the case of a mixed race than in that of a pure line, but is it certain, as assumed by them, that selection is *wholly* without effect in the case of a pure line? We know the effects should be *less*, but are they *nil*? Concerning this matter we are perhaps justified in awaiting further evidence."[c]

East, Castle's colleague at Harvard, felt that the new work on multiple factors had revealed the mischief caused by the unit-character concept. He advocated its abolition. "If one describes a unit character as the somatic expression of a single genetic factor or hereditary unit, he at once gets into trouble. As the factor and not the character is the descriptive unit, a unit factor may affect a character but that character may never be expressed except when several units cooperate in ontogeny. I should prefer to disregard the word character therefore in formulating the problem."[119] But in discarding the unit-character, East substituted the gene in its undefined form. To East, the gene was only a concept "completely free of any hypotheses" as Johannsen had stated, and void of any physical reality. "As I understand Mendelism, it is a concept pure and simple. One crosses various animals or plants and records the results. With the duplication of experiments under comparatively constant environments these results recur with sufficient definiteness to justify the use of a notation in which theoretical genes located in the germ cells replace actual somatic characters found by experiment. . . . Mendelism is therefore just such a conceptual notation as is used in algebra or in chemistry. No one objects to expressing a circle as $x^2 + y^2 = r^2$. No one objects to saying that $BaCl_2 + H_2SO_4 = BaSO_4 + 2 HCl$. No one should object to saying that $DR + RR = 1 DR + 1 RR$."[119] If this notational system were permissible, then "a factor, not being a biological reality but a descriptive term, must be fixed and unchangeable. If it were otherwise it would present no points of advantage in describing varying characters."[119]

It seemed incredible to Castle that anything in a living system could be considered changeless. How could geneticists be dealing with nothing but mere abstractions? This was just Platonism in genetic guise! "In our formulae A is always A, and B is always B, but it is an open question whether in our living animals the characteristics or quantities designated by these symbols are from generation to generation as constant and changeless as the symbols. Bateson and Johannsen and Jennings have assumed that they are. . . ."[68] To look upon changeless formulae as representing changing observable realities was ludicrous and it was impossible for him to accept this view. "The fundamental point of difference between these two views lies in their different conception of unit-characters. To the mutationist unit-characters are as changeless as atoms and as uniform as the capacity of a quart measure. Theoretically an atom is an atom under all circumstances, and a quart holds the same anywhere and everywhere. But the worldly-wise know that the actual quart is not the same in all places; it is apt to be smaller at the corner grocery than in the U.S. Bureau of Standards, and the dishonest tradesman will select effectively for diminished size among the various quart measures offered on the market, unless his selection is carefully restrained by legislation. Similarly, actual unit-characters are modifiable under selection; only one blindly devoted to contrary theory will be able long to shut his eyes to this fact. For several years I have been engaged in attempts to modify unit-characters of various sorts by selection and in every case I have met with success."[68]

Castle was not arguing on the basis of philosophy alone. He had investigated variation in five different allelic pairs of unit-characters— hoodedness in rats; partial albinism in rabbits; coat texture, silver coat coloring, and polydactylism in guinea-pigs. In all cases he could claim the same results. "In my experience every unit-character is subject to quantitative variation, that is, its expression in the body varies, and it is clear that these variations have a germinal basis because they are inherited. By selection plus or minus through a series of generations we can intensify or diminish the expression of a character, that is, we can modify the character. . . . It is the substantial integrity of a quantitative variation from cell-generation to cell-generation that constitutes the basis of Mendelism. All else is imaginary."[68]

These experiments now gave him an opportunity to attempt to dispose of the rival interpretations. "We can distinguish and trace the history of these quantitative variations from generation to generation only when the differences between them are of some size. This has led many to think that only variations of some size are inherited (the mutation theory) and others to deny that such variations can be increased in size by selection (the genotype theory). Others still observing unmistakable evidence that small variations are heritable no less than large ones conceive that the large variations which can be increased or decreased by selection are composed of a certain number of smaller ones cumulative in their effect (the multiple factor hypothesis). A fatal objection to this idea is the fact that these quantitative variations behave as *simple* units, not as multiple ones, and so give monohybrid ratios, not polyhybrid ones. The only logical escape from this dilemma for one devoted beyond recall to a pure-line hypothesis will be to assume further that the assumed multiple units are all coupled; i.e., all united in a single material body so that in cell division they *behave* as one unit. This position will be *logically* unassailable, for we shall never know whether the body which in practice behaves as one is in the last analysis composite."[69] Such a composite view was only a straw man advanced to provide Castle with an *ad hoc* assumption and Castle was firm in invoking Occam's razor to reduce the assumptions to a minimum. His was still the interpretation which required the fewest assumptions!

Bateson, Castle, Johannsen, and East all worked with units which resided in the gametes. Their association with a material object other than the cell itself was purely speculative. But at Columbia University there arose a rival school of genetics. Ironically, the organism they used was supplied to them by Castle: the fruit fly, *Drosophila melanogaster*. On the basis of experiments with *Drosophila*, they advocated a factorial hypothesis (see Chapter 6). The factors of heredity, they claimed, were in a line! Far from being imaginary, these factors were claimed to be parts of chromosomes. The development of this factorial hypothesis as a chromosome theory of heredity was led by Thomas Hunt Morgan and his students. By 1914 they had made several astounding discoveries and Morgan could claim that "a factor, as I conceive it, is some minute particle of a chromosome whose presence in the cell influences the physiological processes that go on in the cell. Such a factor is supposed to be one element only in producing characters of the body. All the rest of the cell or much of it (including the inherited cytoplasm) may take part in producing the characters. So far as such things as unit characters exist I look upon them merely as the most conspicuous result of the activity of some part of the chromosome. A single factor may affect all parts of the body visibly, or a factor may preponderantly influence only a limited section of the body. As a matter of fact, if we look carefully, we can generally find far-reaching effects of single factors."[204]

In contrast to East's view that Mendelian notation was a purely conceptual description of the facts from breeding analysis, Morgan favored a material interpretation of the Mendelian factors. "If one objects to locating these points in chromosomes and prefers to treat biological problems in terms of mathematics he can make the same predictions from the data that can be treated without regard to the mechanism of the chromosomes. But since we find in the chromosomes all the machinery actually at hand for carrying out this procedure, it seems to me reasonable to base our conceptions on this mechanism until another is forthcoming. And if it should prove true that we have found the actual mechanism in the organism that accounts for segregation, assortment, and linkage of hereditary factors we have made a distinct advance in our study of the constitution of the germ plasm."[204]

Castle had kept his fight limited to those who had advocated multiple factors. Morgan and his students now joined this opposition. Castle, however, ignored the first intrusion of the "chromosome theory" and focused his attack on the imaginary quality of the factors. "The biologist's 'pure line' is an imaginary thing. I doubt very much whether it was ever realized in any actual race of animals or plants. It has no more relation to actual animals and plants than a mathematical circle has to the circles described by the most accomplished draftsmen. . . . Since the supposed 'factors' of inheritance are invisible, we cannot hope to deal with them directly by experiment, but only indirectly. Our method obviously should be to eliminate all environmental factors so far as possible and also all factors of inheritance except one. If then this one being present gives a uniform result and being absent gives a result also uniform but different, we can conclude the factor constant. But it is very difficult to apply this method to specific cases, since when variation is observed it is always possible to suppose that all factors but one have not yet been eliminated."[72] Johannsen's work, he charged, had to be irrelevant; "the bearing of . . . Johannsen's . . . observations on seed size upon the question of the constancy of Mendelizing characters is not very obvious, since seed size does not Mendelize. . . . To interpret such cases as Mendelian requires the assumption that no single unit or factor is concerned in size difference, but many wholly independent units."[72] Once again, Castle had to point out that this type of reasoning with superfluous assumptions was not good science; it called for a renewed application of Occam's razor. "In all cases studied critically with reference to the constancy of characters demonstrably Mendelian, the characters have been found to be *inconstant* and subject to modification by selection. What ground is there, then, for supposing that in a case where no factors are demonstrable, such factors are *invariable*? This is like supposing that the moon is made of cheese and that further this cheese is green. The speculation is harmless, if one chooses to amuse himself that way, but it can scarcely be called a valid scientific conclusion."[72]

The unit-character fallacy was now firmly entrenched in genetics. Neither side would concede to the other, and five more years would be needed before its resolution.

CHAPTER 5

The Demise of the Unit-Character

With the advent of World War I, in 1914, the controversy between Castle and his critics seemed to take on a more furious exchange. Castle and Phillips published the results of six years of experimentation with one strain of rats. The character that they chose was a Mendelian recessive—hooded or piebald coat pattern. Beginning with a single strain of these hooded rats, they followed selection in two directions, increasing pigmentation (giving rise to a "plus" series) and decreasing pigmentation (establishing a "minus" series). For thirteen generations they selected in each of these series for the most extreme grades possible. The litters derived from each pair of extreme plus or extreme minus parents were graded in each generation. For both series a continuous modification of the character was maintained. The plus series became darker, the minus series, lighter. The process could be reversed by "return selection" and the change was as slow on the return as it was in becoming established. When Castle outcrossed his homozygous minus lines with wild rats, the "extracted" or F_2 progeny showed a sharp regression to the original hooded condition, just as if a contamination had taken place during heterozygous association of the "minus" hooded allele and its normal allele.

All was in agreement with Castle's view until the complementary test was carried out. An extreme plus grade of homozygous hooded was crossed with the wild-type rat. If contamination were to occur the darker, normal allele should contaminate the plus selected hooded and make it somewhat darker than it had been in the "extracted" or F_2 hooded homozygotes.

-2   -1   0   +1   +2   +3   +4

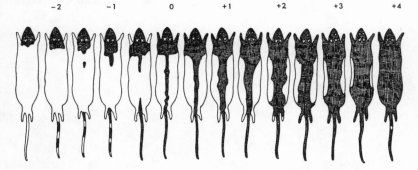

Figure 5.   Castle's Hooded Rats and Gametic Purity. The various + and — grades which Castle used in his selection experiments run from nearly albino to nearly full colored. Selection in either direction affected the mean grade of phenotype in subsequent generations. Castle believed the Hooded factor was altered and selection established the degree of hoodedness. Morgan, East, and others accepted "residual inheritance" or multiple factors as the mechanism for the phenomenon.

But the F_2 hooded rats were not *more* pigmented but *less!* The contamination did not take place. What changed the direction of variation in the "extracted" plus strain? If it were a modifier, another factor affecting the hooded character, this would be a step in the direction of multiple factors. Castle had no alternative; he compromised. "Now it seems to us probable that what we call the unit-character for hooded pattern is itself variable; also that 'modifiers' exist—that is, the extent of the hooded pattern is not exclusively controlled by a single localized portion of the germ-cell; otherwise we should be at a loss for an explanation of the peculiar results from crossing plus series hooded rats with those which are still more extensively pigmented; for by such crosses the pigmentation is rendered not *more* extensive but *less* so. This result we can only explain on the supposition that the selected plus series has accumulated more modifiers of the hooded pattern than the wild race contains, so that a cross tends to reduce the number of modifiers in the extracted hooded individuals. No other explanation at present offers itself for this wholly unexpected but indubitable result. If a different one can be found, we are quite ready to discard the hypothetical modifiers as a needless complication. . . ."[83]

Morgan's group was now in the ascendant and the persistence of Castle's attacks on the factorial hypothesis needed a reply from them. The first encounter involved a critique of Castle and Phillips's paper. Its author was a twenty-two-year-old graduate student, Hermann J. Muller. He criticized the need to maintain factor variation when modifiers were demonstrably present. "Certain crosses proved that more than one factor affecting hoodedness is involved in the difference between the different races. Therefore the production of animals of desired grade by selection may perhaps be explained as a mere sorting out, into different lines of descent, of different combinations of the various factors for hoodedness originally present in the heterozygous ancestors."[218] The basis for Castle's results on this interpretation resided in his original strain, thirteen generations before. "The strain of hooded rats . . . was probably a hybrid between two races of rather remote relationship. When two such races are crossed the individuals often differ in more than one pair of those factors that affect the character studied, especially if the character is such as to be influenced by a relatively large number of genes."[218] The use of the multiple factor hypothesis and the "constancy of a great many genes, 'at one blow' . . ."[218] in any one of Johannsen's pure lines were sufficient to accommodate the data without additionally invoking an instability in the factor for hooded itself. Occam's razor was proving itself to be a two-edged sword! Substituting the "fluctuating or frequent and progressive variation of a gene" was a mistake. Indeed, "it is difficult to believe that this suggestion of Castle and Phillips was not made in a spirit of mysticism, when we consider also their suggestion that the gene may undergo contamination. . . ."[218]

Castle quickly defended his views. In an article submitted to the *American Naturalist* that same year with a title singling out his critic "Mr. Muller on the Constancy of Mendelian Factors," Castle rejected the use of Johannsen's work. "It is difficult to understand how the experiments of Johannsen have any direct bearing on the case since no single Mendelizing unit-factor was demonstrated in that connection; but in the hooded pattern of rats a Mendelizing unit-factor is unmistakably present and it is the quantitative variation of this which is under discussion, not the presence of many or a few additional factors, concerning which Muller adopts our explanation."[70] In his criticism of the use of the gene concept for the hooded series, Castle reveals the unit-character fallacy which has forced him into a corner. Quoting Muller, "In no known case do the variations of a gene among, let us say, several thousand immediate descendants of the individual possessing it, form a probability curve . . ."[218] Castle replies, "The use of the word 'gene' in this sweeping statement safeguards the author, since no one, so far as I know, claims ever to have seen a 'gene' or to have measured it. How could the 'variations of a gene' be expected to 'form a probability curve' if the gene is not measurable? But if the author will allow the substitution of *visible character* for 'gene' in his challenge, I will gladly accept it. . . ."[70]

Although Castle had to back-track on the possibility of gametic contamination as the basis of his outcrossed series, he now became its defender. "Mr. Muller is seriously disturbed . . . because we are willing to consider it possible that the 'factor for hooded' may be contaminated by 'its allelomorph' while associated with it in the zygote represented by the F_1 rats. . . . He says this is 'violating one of the most fundamental principles of genetics—the non-mixing of factors—in order to support a violation of another fundamental principle—the constancy of factors.' Now, when, I should like to inquire, did these principles become 'fundamental,' by whom were they established and on what evidence do they rest? . . . We shall look in vain, I think, for those 'principles' outside of the 'Exakten Erblich-keitslehre' (or its imitations), and when we inquire as to the experimental basis of the principle in question we are met with the satisfied reply, 'Johannsen's beans.' What a slender basis and what an absurd one from which to derive the 'fundamental principles' that Mendelian factors are constant! . . . Do biologists take themselves seriously when they reason thus? Certainly no one else will long take them seriously."[70]

In another article that year Castle took a firm position on the multiple factor hypothesis. "If the multiple factor hypothesis must stand or fall with the pure line doctrine, I for one cannot accept it, for the foundations of the pure line doctrine appear to me very insecure."[72] East's interpretation of the multiple factor theory he dismissed as valueless, "It is of course possible as Dr. East maintained to formulate a description of all heredity in terms of (purely subjective) Mendelian units, provided more and more units are from time to time created (by imagination) as the objective facts show the organism changed. But such an extension of Mendelism fails to interest me, as I think it does many of my readers. What we want to get at, if possible, is the objective difference between one germ cell and another, as evidenced by its effect upon the zygote, and it is the constancy or incon-stancy of these objective differences that I am discussing. If these are quantitatively changeable from generation to generation, then change in the variability of the zygotes composing a generation might arise without factorial recombination."[71]

The following year, Castle announced that he had a critical experiment bearing on the "objective difference" between germ cells. "As a crucial experiment, we conceived the plan of deriving an entire race of animals, not from a single *pair* of ancestors, but from a single gamete, so far as concerns a particular unit-character. It was thought that in a race so derived, if the principle of gametic purity holds, there should be no variation whatever in the particular unit-character concerned. . . ."[81] The unit-character which Castle, with his student Hadley, had pursued was the English spotted pat-tern in rabbits. Despite the origination from a single gamete, selection produced different grades of spotting. "Hence the English unit-character has changed quantitatively in transmission from father to son. This seems to us conclusive evidence against the idea of unit-character constancy, or 'gametic purity.' If unit-characters are not constant, selection reacquires much of the importance which it was regarded as possessing in Darwin's scheme of evolution, an importance which many have recently denied to it."[81]

Morgan's laboratory summarized the findings of four years research with *Drosophila* and a book-length defense of their chromosome theory was published, also in 1915, as *The Mechanism of Mendelian Heredity*. The Morgan School sharply criticized the unit-character and restored, to its original status in Mendel's paper, the two components of Bateson's concept. "Failure to realize the importance of these two points, namely, that a single factor may have several effects, and that a single character may depend on many factors, has led to much confusion between factors and characters, and at times to the abuse of the term 'unit-character.' It cannot, therefore, be too strongly insisted upon that the real unit in heredity is the factor, while the character is the product of a number of genetic factors and of environmental conditions. The character behaves as a unit only when the contrasted individuals differ in regard to a single genetic factor, and only in this case may it be called a unit character. . . . So much misunderstanding has arisen among geneticists themselves through the careless use of the term 'unit character' that the term deserves the disrepute into which it is falling."[217]

A similar view, expressed by Sturtevant in Morgan's laboratory, reflects the difference in philosophy between the two groups. "The frequency of the use of the term 'unit character' where gene or 'unit factor' is obviously what is meant is an illustration" of the confusion that terminology can bring. "This term 'unit character' is an unfortunate one, as it implies the conception that characters are in some way caused by single genes—a view which is not tenable in the light of our modern knowledge, and is, I feel certain, really held by few, if any, geneticists today. It focuses the attention upon the somatic characters of organisms rather than upon their germinal constitution; whereas the present study of genetics is tending in exactly the opposite direction."[299]

Castle at first scoffed at these rival criticisms. "The idea of fixity among living things seems to be one which the human mind is loath to give up and which has to be constantly combatted in the advancement of biology. For centuries it was the fixity of species which dominated biological thought. Darwin had to dispel this idea before he could get a hearing for evolution. When the Mendelian theory of unit-characters came in, the idea of fixity, unchangeableness, attached itself to the unit-characters. Driven from this hold, it now seizes on the single factors on which Mendelian characters depend. Simultaneously, it attaches itself to the conjectural mechanism which underlies Mendelian heredity, the chromosomes."[74] It was not long, however, before Castle designed even more critical experiments in his hooded rats to test the strength of his selected plus and minus strains. These results were decisive. In 1919 Castle submitted his paper "Piebald Rats and the Theory of Genes" to the *Proceedings of the National Academy of Sciences.* "My own experimental studies of heredity, begun in 1902, early led me to observe characters which were unmistakably *changed* by crosses and so I have for many years advocated the view that the gametes are not pure in the sense expressed by Bateson. . . . Characters . . . even when uncrossed, show fluctuating or graded variations in consequence of which systematic selection is able to produce very diverse races as regards a single Mendelizing character, the ordinary allelomorph of which is wholly excluded from the experiment. This observation shows that characters may vary otherwise than by contamination and I was in consequence led to adopt the hypothesis that unit-characters are 'inconstant' in varying degrees, but probably never constant. . . . This view has been repeatedly challenged, either by those who questioned the evidence cited in support of it, or by those who substituted a different concept, 'gene,' for that of 'unit-character' and then denied that a 'gene' can vary."[75]

The significance of the gene concept, as Castle understood it was precisely as Johannsen had advocated—it was "the thing in the germ cell which produces the visible character." But "to express heredity in terms of unvarying genes, it is necessary to suppose that besides the single gene indispensible to the production of a visible character, its gene proper, there occur also other genes whose action is subsidiary. . . . These are called modifying genes. . . . In the majority of cases the only grounds for hypothecating the existence of modifying genes is the fact that characters are visibly modified. . . . As an alternative to the theory of modifying genes, the theory has been considered that genes themselves may be variable and if so, genes purely modifying in function might be dispensed with."[75]

When a highly selected plus strain of hooded rats was crossed to wild rats in 1916, "the hooded character was lowered not over three-fourths of a grade by three successive crosses. This fact led me to conclude provisionally in 1916 that the hooded gene proper had really changed in the course of our selection experiments, since after the crosses it *remained different* from what it had been originally."[75] But "this view is obviously erroneous in the light of the results obtained from the minus crosses subsequently studied."[75] The minus strains were obliterated in three generations of outcrossing to wild rats, and jumped to a plus grade comparable to those of selected plus strains! "These differences accordingly were based on residual heredity, not on changes in the hooded gene proper. For when the residual heredity was equalized, the hooded character appeared substantially the same in the two races. These findings harmonize with the idea that the residual heredity in question consists of several modifying genes independent of the hooded gene proper."[75] In concluding, Castle could not avoid a full statement of the significance of his results. "These results favor the widely accepted view that the single gene is not subject to fluctuating variability, but is stable like a chemical compound of definite composition and changes only similarly, by definite steps (mutation in the sense of Morgan, not of de Vries). They offer no obstacles to the proposition of Johannsen (ably supported by East), that a gene terminology is adequate to express all known varieties of inheritance phenomena."[75]

World War I had ended; but for Castle, it was not Armistice; it was unconditional surrender.

CHAPTER 6

Associative Inheritance

A few years after Bateson had left the Chesapeake Bay Biology Station, a twenty-year-old graduate from the University of Kentucky began graduate studies at Johns Hopkins University. Thomas Hunt Morgan had an enthusiasm for natural history and he had heard of the fame of Johns Hopkins when he spent a summer at a marine biology station in Massachusetts. Brooks, with his colleagues, had a good number of promising students. E. B. Wilson had just graduated and was devoting his studies to the cell; E. G. Conklin and R. G. Harrison were beginning their graduate studies along with Morgan. "It was Brooks who influenced him in his choice of embryology as his first field of study . . ."[304] but Morgan felt more attracted to the physiologists and their experimental approaches than he did to Brooks who was ". . . too metaphysical for his tastes."[304] He received his Ph.D. in 1890 for investigations of the embryology of sea spiders (Pycnogonida). Like Bateson, he found the embryology of *Balanoglossus* intriguing and he extended his observations to related hemichordates, especially ascidians. At the turn of the century, all embryology fascinated him, particularly the developmental and physiological problems of fertilization, parthenogenesis, sexuality, and regeneration.

Morgan was skeptical of the highly speculative trends in embryology which used particulate units to explain all differentiation and heredity. He thought these attempts were "preformationist" in outlook. "The idea of preformation has assumed a new form in the hands of modern thinkers. It is assumed by certain modern embryologists, of whom Weismann may be taken as an example, that there exist in the chromatin of the individual certain bodies that correspond to each character of the individual, and that the process of development consists in the sorting out of these bodies to the

39

different parts of the egg."[195] In this same category he placed the theories of Roux. "Roux attempted to show how the qualitative division of the egg is brought about. He pointed out that the spindle that is formed during each cell division, and by means of which the separation of the chromatin takes place, is an instrument by which we can conceive a qualitative distribution of the chromatic material to be accomplished. The chromatin is looked upon as a substance bearing the essential qualities of the individual, as is demonstrated by the inheritance of the characters of the father through the nucleus of the spermatozoan. This same idea of qualitative separation at each division of the qualities of the chromatin is, as I have said, also the central idea of Weismann's hypothesis."[195] As an embryologist, Morgan believed that numerous physiological events participated in the organism's development and that embryonic complexity was of this "epigenetic" nature rather than the particulate "preformationist" systems so widely advocated.

In 1900 Morgan visited the laboratories of Hugo de Vries at Hilversum, in Holland, and became impressed by the numerous discontinuous variations that de Vries had obtained in the evening primrose, *Oenothera lamarckiana*. De Vries, at this time, unlike Bateson, was not advocating Mendelism as the major contribution to studies of evolution; rather, he proposed his own "Mutationstheorie." After his ideas in intracellular pangenesis had taken form, de Vries sought organisms which provided discontinuous variations by "leaps and jumps." Such major departures in an organism's typical characteristics he called *mutations*. "Therefore I have sought for a plant which would produce more of such mutations than other plants. I have studied over a hundred varieties, investigating their progeny, and among them one has answered my hopes. This is the evening primrose of Lamarck, which chances to bear the name of the founder of the theory of evolution which it is prepared to support."[114] De Vries does not imply that acquired characteristics is the method of evolution; his tribute to Lamarck is for his popularization that evolution had occurred; it was Darwin who found "a conclusive proof of the idea of Lamarck."[114]

Mutations were the exact opposite of the mechanism proposed by Lamarck. In *Oenothera* their occurrence was pronounced and dramatic: "In these cultures the species is seen to be very pure and uniform in the large majority of its offspring, but to produce an average one or two aberrant forms in every hundred of its seedlings. The differences are easily seen even in young plants and are mostly large enough to constitute new races. The more common ones of these races are produced repeatedly, from the seed from the wild plants as well as in the pure lines of my cultures. It is obviously a constant and inheritable condition which is the cause of these numerous and repeated jumps. These jumps at once constitute constant and ordinary uniform races, which differ from the original type either by regressive characters or in a progressive way. By means of isolation and artificial fecundation, these races are easily kept during their succeeding generations."[114]

Morgan championed this view. In 1905 he claimed, "The time has come, I think, when we are beginning to see the process of evolution in a new light. Nature makes new species outright. Amongst these new species there will be some that manage to find a place where they may continue to exist. . . . Some of the new forms may be well adapted to certain localities, and will flourish there; others may eke out a precarious existence, because they do not find a place to which they are well suited, and cannot better adapt themselves to the conditions under which they live; and there will be others that can find no place at all in which they can develop, and will not even be able to make a start. From this point of view the process of evolution appears in a more kindly light than when we imagine that success is only attained through the destruction of all rivals. The process appears not so much the result of the destruction of vast numbers of individuals, for the poorly adapted will not be able to make even a beginning. Evolution is not a war of all against all, but it is largely a creation of new types for the unoccupied, or poorly occupied places in nature."[195] Morgan took an interest in problems of heredity; like Castle he criticized the idea of gametic purity. He followed the new chromosomal studies of sexuality which were being developed by his colleague and friend at Columbia University, E. B. Wilson. Wilson, with Morgan's student, Stevens, had demonstrated that the "accessory" or "X" chromosome observed in 1901 by McClung in grasshoppers was actually a single X in males of many insect species and two X's in the females of these species. In trying to pursue these ideas in aphids, Morgan found that the chromosomal mechanism was even more complex. But there was an unquestioned chromosomal association with sexuality.

While Morgan was concentrating his attentions on these cytological problems, Bateson, in 1906, reported a new discovery. "In segregation, features of structure or of physiological constitution are treated as *units* by the cell-divisions in which the germs are formed. Such segregation characters are, as far as we know, always constituted in pairs. . . . Sometimes, however, there is evidence of a linking or *coupling* between distinct characters. When such a coupling is complete, the two characters, of course, can be treated as a single allelomorph. . . . But besides this simpler phenomenon of complete coupling we now know that the usual ratios are liable to disturbance by a *partial coupling* between distinct characters. Such examples have not yet been fully studied."[17] Bateson and his colleague, Punnett, found certain unit-characters in sweet peas which did not give the anticipated $9 : 3 : 3 : 1$ ratios among the F_2 progeny. "Early in the revival of breeding experiments, attention was called, especially by Correns, to the phenomenon of coupling between characters. Complete coupling has so far been most commonly met with among characters of similar nature. Examples of complete coupling between characters apparently quite distinct in nature are less frequent. . . . Examples of *partial* coupling have not hitherto been adequately studied. A remarkable case occurs in regard to the distribution of the pollen characters in F_2 from the white long x white round Sweet Pea. There is here a partial coupling between the purple flower-colour and the long pollen. The whole mass of F_2 consists of 3 long : 1 round. The white taken alone are also 3 long : 1 round. The *purples*, however, show a great preponderance of longs over rounds, about 12 to 1, while among the *reds* the rounds are in excess over the longs, being about 3 to 1. These peculiar distributions only occur in families which contain both red and purple members. . . . Evidently the abnormal distribution in some way depends on the mode of segregation of the factors B and b from each other. A close approach to the observed F_2 numbers would be given by a system in which each 16 gametes were composed thus:—7AB + 1 aB + 1 Ab + 7 ab—where A is long pollen and a round pollen . . ."[28] and B and b represent the allelomorphs for one of the color factors.

Two years later Bateson and his colleagues reported a new case of "partial coupling" giving a different "gametic ratio" and they proposed a series of ratios, geometrically discontinuous. "The undoubted existence of these two grades of gametic coupling in the Sweet Pea suggests that each may find its place in a scheme of increasing intensity of gametic coupling, such as is shown in the accompanying table, where the two allelomorphic pairs are represented by Aa and Bb":[29]

gametic series				zygotes containing				
AB	Ab	aB	ab	A&B	A only	B only	A nor B	
1	1	1	1 = 4	9	3	3	1 =	16
3	1	1	3 = 8	41	7	7	9 =	64
7	1	1	7 = 16	177	15	15	49 =	256
15	1	1	15 = 32	737	31	31	225 =	1024
n-1	1	1	n-1 = 2n	$3n^2 - (2n-1)$:	2n-1 :	2n-1 :	$n^2 - (2n-1)$ =	$4n^2$

Notice, in the system proposed by Bateson, that "the closer the gametic coupling the rarer become the two middle terms which contain but one dominant apiece."[29] This distinction between a gametic and zygotic system was based on an extrapolation from the F_2 progeny to the inferred ratios of F_1 gametes, bred *inter se*, which would have given such a frequency of F_2 progeny.

Additionally, another phenomenon appeared in 1908 which departed from the typical Mendelian ratios. "In the cases of actual gametic coupling . . . the association is always between the dominant or *present* factors of the different pairs. . . . The 'present' factors would seem to behave in gametogenesis as though they were attracted to one another. But here the phenomenon is of a different order. The two dominant factors . . . seem to repel one another so that they are not both found in the same gamete. . . . For this phenomenon we suggest the term *spurious allelomorphism*."[29] The choice of this term was based on the resemblance to true allelism, in which a pair of alternative factors always segregated from the hybrid; in this case two unrelated factors appeared to segregate at all times from the hybrid. This term was changed shortly afterwards, however, to repulsion.

While "coupling and repulsion" were exciting the attention of European geneticists, Morgan was still not convinced that the chromosomes were directly related to the problem of heredity. "We have come to look upon the problem of heredity as identical with the problem of development. The word heredity stands for those properties of the germ cells that find their expression in the developing and developed organism."[198] Speculation linking the chromosomes to Mendelian factors had already been made. "Since the number of chromosomes is relatively small and the characters of the individual are very numerous, it follows on the theory that many characters must be contained in the same chromosome. Consequently many characters must Mendelize together. Do the facts conform to this requisite of the hypothesis? It seems to me that they do not."[198] Morgan was criticizing the view suggested in 1902 by W. Sutton that chromosome behavior at the time of gamete formation (meiosis) paralleled that of Mendelian unit-characters. "If Mendelian characters are due to the presence and absence of a specific chromosome, as Sutton's hypothesis assumes, how can we account for the fact that the tissues and organs of an animal differ from each other when they all contain the same chromosomal complex? Bateson has called attention to this weakness of the single-chromosome : single-character hypothesis. . . ."[198] Early in 1910 Morgan confessed that he had an agnostic attitude to the chromosomal basis of heredity, "I have tried . . . to weigh the evidence, as it stands, in the spirit of the judge rather than in that of the advocate."[198]

About six months earlier, Morgan was using fruit flies in his laboratory as a new organism for his researches. Castle had shown that they were convenient to handle and that they grew in large numbers at a minimal expense. Morgan hoped that such a fast breeding system (two to three weeks per generation) would be useful for detecting induced mutations. With this idea in mind he tried a variety of agents, particularly different salts, radium, and x-rays, but nothing like de Vries' phenomenal mutations occurred. In one of his pedigreed cultures, however, a single white-eyed male appeared in a normal population of red-eyed flies. Breeding this to its sisters, Morgan found that the F_1 progeny were red-eyed; this result would make the white eye coloration recessive. The F_2 gave a 3 : 1 ratio, but all the white-eyed flies were males and the red-eyed flies were in a ratio of two females to one male. A second cross of white-eyed males to F_1 red-eyed females gave both red- and white-eyed progeny of both sexes in equal numbers. "The results just described can be accounted for by the following hypothesis. Assume that all of the spermatozoa of the white-eyed male carry the 'factor' for white-eyes 'W,' that half of the spermatozoa carry a sex factor 'X,' the other half lack it, i.e., the male is heterozygous for sex. Thus the symbol for the male is 'WWX,' and for his two kinds of spermatozoa WX-W. Assume that all of the eggs of the red-eyed female carry the red-eyed 'factor' R; and that

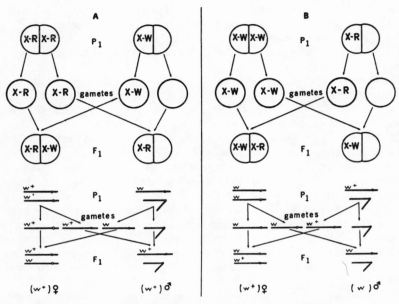

Figure 6. Symbolic Representation of Sex-linked Inheritance. Morgan used Stevens and Wilson's original observation that *Drosophila* females contained two sex factors (XX) and the males only one (XO). The factor for red eyes (R) was associated with the sex factor X. In the white mutation this factor was originally assumed to be present in both the X and the O. Morgan decided, however, that the W factor was only present in the X because normal males had neither R nor W in their O sex factor. Morgan also waited for more mutations to occur before he associated these sex factors with the X chromosome. In A, the top figure illustrates Morgan's cross of red-eyed females with white-eyed males. In B, the reciprocal cross of white-eyed females and red-eyed males yields a 1 : 1 ratio of red-eyed females and white-eyed males. This reversal of sex and phenotype was called "criss-cross" inheritance. The chromosomal representation, with X and Y chromosomes and the alleles w+ (red) and w (white), are shown in the lower figures.

all of the eggs (after reduction) carry one X each, the symbol for the red-eyed female will be therefore RRXX and that for her eggs will be RX-RX."[199]

Using this terminology, Morgan assumed his first found white-eyed male was homozygous for W, but heterozygous for the sex factor X. Thus normal, unrelated strains of red-eyed fruit flies should be homozygous for R and heterozygous for X, giving them the composition RRX. Such a red-eyed male crossed to one of the white-eyed females should then give all red-eyed progeny in the F₁. But "a most surprising fact appeared when a white eyed female was paired to a wild, red eyed male, i.e., to an individual of unrelated stock. The anticipation was that wild males and females alike carry the factor for red eyes, but the experiments showed that all wild males are heterozygous for red eyes and that all the wild females are homozygous.

Thus when the white eyed female is crossed with a wild red eyed male, all of the female offspring are red eyed, and all of the male offspring are white eyed."[199] From this Morgan concluded that the non-mutant strains, the original wild type, red-eyed males are all heterozygous for W! But if all red-eyed males were heterozygous for white, then why didn't homozygous white-eyed *females* appear relatively frequently in the *wild* population? Morgan had to make the further assumption that the X factor was coupled with R. This would keep all males red (in effect, permanently heterozygous as RX-W) and all females red because they always had the X factors (RX-RX). The mutation (or sport, as Morgan called it) then had to occur in one of the *eggs* of the red-eyed females; it must have changed from RX to WX.

Bateson had suggested that white usually represented the loss or absence of a factor for color. Using this prevailing notion, Morgan discarded his former symbolism, "It seems probable that the sport arose from a change in a single egg of such a sort that instead of being RX (after reduction) the red factor dropped out, so that RX became WX or simply OX."[199] The normal strains would then be RRXX females and RXO males. The white-eyed females were OOXX and the white-eyed males, OOX. "It now becomes evident why we found it necessary to assume a coupling of R and X in one of the spermatozoa of the red F_1 hybrid (RXO). The fact is that this R and X are combined, and have never existed apart."[199]

Although this point was the most significant finding in his experiment, Morgan was more impressed by a different implication of his work. In England Doncaster, Bateson, and Punnett had found numerous instances of the type of "sex-linked" inheritance which Morgan had found in *Drosophila*. The moth *Abraxas*, canaries, and fowl all showed that such "sex-limited" traits were heterozygous in the female and homozygous in the male. "The most important consideration from these results is that in every point they furnish the converse evidence from that given by *Abraxas* as worked out by Punnett and Raynor. . . . Significant, too, is the fact that analysis of the result shows that the wild female *Abraxas grossulariata* is heterozygous for these two characters."[199]

It is significant that Morgan did not state that this case of "sex-limited" inheritance was associated with a chromosome. Throughout, Morgan maintains his agnosticism and uses the ambiguous phrase "sex factor" or "X." This probably reflected the realization, at the time of his discovery, that the cytological work in the United States did not agree with the genetic work in Great Britain. Wilson and Stevens showed convincingly that males were bearers of a single "accessory" or X chromosome which the females had as a pair in the species they investigated. The British work, in contrast, showed that the *genetics* of *Abraxas* and domesticated birds was exactly the opposite! The white-eyed case was in agreement with the cytological work, but was one case sufficient proof of an association between sex-limited inheritance and the X chromosome?

Morgan found more mutants in 1910 and two of these, yellow body color and miniature wings, were also "sex-limited." There was no question in Morgan's mind that the correlation of these factors with sex demanded their presence in the single X chromosome. But when testing this, Morgan found that the coupling was not complete! A recombination of factors resulted from crosses of white eyes with miniature wings and of white eyes with yellow body color. Now Morgan could cast off his agnosticism and assert a novel interpretation of heredity which unified two independent lines of approach. "In place of attractions, repulsions, and orders of precedence, and the elaborate systems of coupling, I venture to suggest a comparatively simple explanation based on results of inheritance of eye color, body color, wing mutations and the sex factor for femaleness in *Drosophila*. If the materials that represent these factors are contained in the chromosomes, and if these factors that 'couple' be near together in a linear series, then when the parental pairs (in the heterozygotes) conjugate, like regions will stand opposed. There is good evidence to support the view that during the strepsinema stage homologous chromosomes twist around each other, but when the chromosomes separate (split) the split is in a single plane,

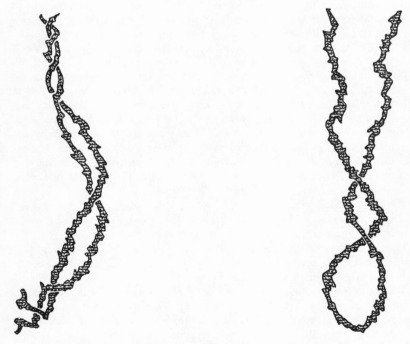

Figure 7. Janssens's Chiasmatype Figures. These illustrations, from Janssens's *La Cellule*, provided the stimulus for Morgan's theory of crossing over. Note that Janssens visualized the chromosomes as single chromatids rather than as double chromatids.

as maintained by Janssens. In consequence the original materials will, for short distances, be more likely to fall on the same side of the split, while remoter regions will be as likely to fall on the same side as the last, as on the opposite side. In consequence we find coupling in certain characters, and little or no evidence at all of coupling in other characters; the difference depending on the linear distance apart of the chromosomal materials that represent the factors. . . . The results are a simple mechanical result of the location of the materials in the chromosomes, and of the method of union of homologous chromosomes, and the proportions that result are not so much the expression of a numerical system as of the relative location of the factors in the chromosomes. Instead of random segregation in Mendel's sense we find 'Associations of factors' that are located near together in the chromosomes. Cytology furnishes the mechanism that the evidence demands."[201]

Morgan's cytological interpretation of exchange was based on Janssens's observations of twisting (chiasmata) among paired chromosomes during the early stages of meiosis. This was called the chiasmatype theory. Morgan adopted it as the physical basis for the association and recombination of factors. "The important point is that the coupling (association) of sex-limited characters that I have found in Drosophila shows that the factors must be referred to the same chromosome, and if so there seems to be no

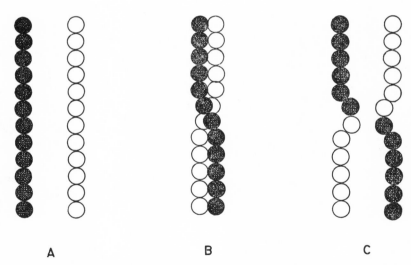

A   B   C

Figure 8.   Morgan's Theory of Crossing Over. The illustration, from 1915, represents the oversimplified "beads on a string" representation of the factors (the word gene is avoided in Morgan's early treatment of crossing over). The chromosome is visualized as containing single chromatids. A represents the parental chromosomes. B represents the crossover based on a cytological chiasma; and C represents the recombinant crossovers. (After Morgan, Proc. Nat. Acad. Sci. 1:420-429.)

escape from the conclusion that the interchange as well as association must be admitted on the chromosome hypothesis."[200]

A more speculative interpretation of chromosome exchange, based on the later reflections of "intracellular pangenesis," was offered, without experimental support, in 1903 by de Vries. "It is another question whether the individual chromosomes correspond also to special groups of hereditary characters, or, in other words, whether the bearers of the latter are strictly localized in the nuclear threads. . . . If every unit, that is, every inner character or every material bearer of an external peculiarity, forms an entity in each pronucleus, and if the two like units lie opposite each other at any given moment, we may assume a simple exchange of them. . . . How many and which, may then simply be left to chance. In this way all kinds of new combinations of paternal and maternal units may occur in the two pronuclei, and when these separate at the formation of the sexual cells, each of them will harbor in part paternal, in part maternal units."[112] Such an interpretation suffered from a lack of predictive value; in contrast, both Morgan and Bateson could predict the experimental consequences of "associative inheritance" or "coupling."

CHAPTER 7

Crossing Over Versus Reduplication

Bateson looked upon the normal, dominant character of an organism as the expression of some unit actually present in the germ cell which introduced it into the zygote. The recessive trait was often considered to be a loss of the dominant character (in color, for example, partial or total albinism, as a general rule, was recessive to its normally colored counterpart). Bateson considered that such a loss must arise, in turn, from the loss or absence of the factor normally present in the cell. When complete coupling was found between dissimilar dominant characters, the attraction between these unit-characters was considered to be at a maximum. Partial coupling suggested an impairment in the attractive forces which coupled the dominant factors. For complete coupling, in a heterozygote AaBb, whose parental gametes were AB and ab, the F_2 progeny would be in a ratio of 3 AB : 1 ab. Partial coupling would introduce two additional classes of progeny, Ab and aB. The frequency of these additional classes would depend on the strength of coupling between A and B. If the coupling was weak, then a gametic ratio of 3 AB : 1 Ab : 1 aB : 3 ab might arise. This would give a zygotic ratio (by squaring the gametic ratio) of 41 AB : 7 Ab : 7 aB : 9 ab. If the coupling was very strong, with a gametic ratio of 63 AB : 1 Ab : 1 aB : 63 ab, the resulting zygotic ratio would be in a ratio of 12,161 AB : 127 Ab : 127 aB : 3969 ab. Strong coupling, in other words, would give almost a 3 : 1 ratio of parental (coupled) classes and the two classes of single dominants would be rare in the population.

In contrast to coupling, spurious allelomorphism, or repulsion, was based on the assumption that the two unrelated, dominant, factors always kept apart from each other in the formation of germ cells. If C and D are the dominant factors, then the hybrid, CcDd, from the union of the repulsed gametes Cd and cD, would give progeny of three types. In addition to the parental forms, expressing cD and Cd, there would be numerous heterozygotes expressing CD (the composition of these CD individuals would, of course, remain in a state of repulsion). The F_2 or zygotic ratio would be 2 CD : 1 cD : 1 Cd.

In 1911, Bateson and Punnett found an exception to this 2 : 1 : 1 repulsion ratio. "If . . . the heterozygote, AaBb, is formed by the gametes Ab and aB, repulsion occurs between A and B, so that only the two classes of gametes Ab and aB are formed. In the account to which we have alluded we supposed that such repulsion was complete, and that the two classes of gamete AB and ab were not formed. Our work on sweet peas during the present summer has led us to modify our conception on the nature of the gametes produced in cases where repulsion occurs. . . ."[25] In a repulsion cross of the same pollen shape and flower color characters in sweet peas that they had used several years earlier for their analysis of coupling, Bateson and Punnett obtained 226 blue long (BL); 95 blue round (Bl); 97 red long (bL); and 1 red round (bl) plants. The obvious classes were in a 2 : 1 : 1 ratio, but where did the doubly recessive, red and round, plant come from? "Either we must look upon it as an unaccountable mutation, or we must consider that the repulsion between B and L is partial. In the light of the evidence afforded . . . we prefer the latter hypothesis, and we are inclined to regard the partial repulsion between B and L as of the 1 : 7 : 7 : 1 type. . . . It is worthy of note that the coupling between B and L is usually on the 7 : 1 : 1 : 7 system, and it would be interesting if in such cases as these the repulsion and coupling systems for a given pair of factors were shewn to be of the same intensity."[25] Repulsion was not only partial; it was the inverse of coupling.

From additional cases, and by deduction from their coupling series, Bateson and Punnett drew up a table of repulsion ratios using the same geometric series as they had for their coupling series.

gametic series				zygotes containing			
AB	Ab	aB	ab	A&B	A only	B only	A nor B
1	1	1	1 = 4	9	3	3	1 = 16
1	3	3	1 = 8	33	15	15	1 = 64
1	7	7	1 = 16	129	63	63	1 = 256
1	15	15	1 = 32	513	255	255	1 = 1024
1	(n-1)	(n-1)	1 = 2n	$2n^2 + 1$	n^2-1	n^2-1	1 = $4n^2$

The reason for a five-year delay between the discovery of coupling and the discovery of partial repulsion is obvious from an inspection of the zygotic ratios for partial coupling and partial repulsion. The double recessive class, *ab*, which proves the occurrence of partial repulsion, is so rare, even when the force of repulsion is weak, that large numbers of progeny would be required to detect it. Partial coupling, on the other hand, permits the recognition of single dominant classes in all except the most intense association of dominant factors.

Bateson was well aware of the cytological studies of meiosis which showed that only a few (usually two) divisions accomplished the process of gamete formation from immature germ cells. Some of the geometric series inferred from actual cases of partial coupling could not be accomplished with less than four cell divisions. Without some sort of model based on cell division, Bateson could not account for the sorting of the gametes into the ratios of the different series. Obviously, meiosis could provide no mechanism for these series; "no simple system of dichotomies could bring about these numbers . . . it was scarcely possible that such a series could be constituted in the process of gametogenesis of a plant, in whatever manner the divisions took place."[25] With meiosis ruled out, Bateson had to infer that "segregation may occur earlier than gametogenesis."[25]

Coupling and repulsion needed a mechanism. The terms were only descriptive of the states of association of the factors present in a germ cell. The geometrical series had to be a consequence of that mechanism and they had to give rise to these two phenomena. "In view of what we now know, it is obvious that the terms 'coupling' and 'repulsion' are misnomers. 'Coupling' was first introduced to denote the association of special factors, while 'repulsion' was used to describe dissociation of special factors. Now that both phenomena are seen to be caused not by any association or dissociation, but by the development of certain cells in excess, those expressions must lapse. It is likely that terms indicative of differential multiplication or proliferations will be most appropriate. At the present stage of the inquiry we hesitate to suggest such terms, but the various systems may conveniently be referred to as examples of *reduplication*, by whatever means the numerical composition of the gametes may be produced."[25]

The *reduplication hypothesis* could now be cast into model form. "At some stage in the embryonic development or perhaps in later apical divisions we can suppose that the n-1 cells of the parental constitution are formed by successive periclinal and anticlinal divisions of the original quadrants which occupy corresponding positions." As an afterthought on the significance of reduplication, Bateson found an evolutionary curiosity which would have to be tested, "Hitherto, no case of *coupling* has been found in animals. . . . At present it seems not impossible that the two forms of life are really distinguished from each other in these respects."[25]

But an animal had been found with partial coupling! It was being tested with almost feverish excitement 3000 miles away in Schermerhorn Hall at Columbia University. Morgan increased the scope of his work and added Calvin B. Bridges and Alfred H. Sturtevant to his laboratory staff. "In the winter of 1910–11 Morgan took C. B. Bridges and the writer—both

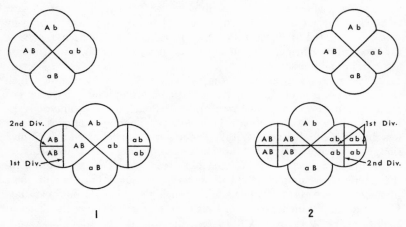

Figure 9. The Reduplication Hypothesis. The non-Mendelian ratios obtained in crosses involving two pairs of characters were interpreted incorrectly by Bateson and Punnett. They assumed a differential multiplication of cells (e.g., AB) leading to fixed geometric ratios (in 1, 3AB : 3ab : 1Ab : 1aB; in 2, 4AB : 4ab : 1Ab : 1aB). These replications had to be mitotic, not meiotic, and they were restricted to germinal cells. (After Punnett, 1913, J. Genet. 3:77-103.)

then undergraduates—into his laboratory, and gave us desks in what came to be known as the 'fly room.' This was a rather small room, with eight desks crowded into it, in which the three of us reared *Drosophila* for the next seventeen years."[304] The "associative inheritance" of 1911 was now more advanced. "In recent years a new fact in Mendelian Inheritance has come to light, which while it obscures the Mendelian expectation based on *independent* segregation of the factors of inheritance, shows that the main Mendelian principles are by no means invalidated; for they too are manifest, but obscured by the linkage or coupling of certain factors. When certain somatic characters are associated with sex, as in Drosophila, we have the best opportunity, as yet afforded, for studying in its simplest form this sort of 'associative' inheritance; for, in certain combinations, the relation between linkage and breakage of the linkage ('crossing-over,' as we shall call it) is shown at once by the male flies which indicate without complications the kinds of eggs that the F₁ female produces."[216] There was no delay in discovering both "coupling and repulsion" in *Drosophila*; they were in fact discovered almost simultaneously, because the F₂ males were effectively haploid for the X chromosome and thus the zygotic male ratio was identical to the gametic ratio. "Partial *repulsion*" was actually discovered before coupling in the cross of white-eyed flies to those with miniature wings.

In November, 1912, Sturtevant made use of Morgan's suggestion that the strength of linkage was dependent "on the linear distance apart of the chromosomal materials that represent factors." Accepting this, Sturtevant advanced a quantitative scheme for converting the frequencies of crossovers

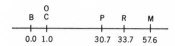

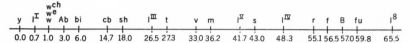

Figure 10.   The Evolution of the Genetic Map. Sturtevant attempted the first map of six sex-linked genes. The symbols he used in 1913 were changed in 1915. The locations of these mutations today are shown in the middle map. These locations vary little from those first determined by Sturtevant. By 1915 there was a great increase in detailed localizations in the X chromosome. In Sturtevant's map y = yellow body color; w and w^e = white and eosin; v = vermilion eye color; m = miniature wing; and r = rudimentary wing. In the original symbolism B = body color factor; C and O the eye color factors involved in the white eye color factors; P = the normal pink eye factor carried by the vermilion mutant; R = the normal rudimentary factor carried by the miniature mutant; M = the normal miniature factor carried by the rudimentary factor. Since 1915 more than 100 genes have been localized to unique sites in the X chromosome.

into a meaningful model. "It would seem, if this hypothesis be correct, that the proportion of 'cross-overs' could be used as an index of the distance between the two factors. Then by determining the distances (in the above sense) between A and B and between B and C, one should be able to predict AC. For, if proportion of cross-overs really represents distance, AC must be approximately, either AB plus BC, or AB minus BC, and not any intermediate value."[297] Sturtevant's model provided him with a basis for mapping the chromosomes. The percents of crossing over that Morgan and others had calculated for the different mutants in *Drosophila*, could now be considered representative of distances along the map. Each percent of crossing over was a unit on the map.

At this time there were six "sex-limited" factors, all associated with the X chromosome and cast into a single map by Sturtevant. In 1912, Morgan and Lynch found another group of mutants which were not sex-linked and which recombined with one another; the mutants were black body color and vestigial wings. "These were considered as lying in the 'second chromosome.' . . . (The) existence of still another group of genes, which are located in the 'third chromosome' . . ."[296] was demonstrated by Sturtevant.

Sturtevant's analysis points out the similarity between Bateson's use of plant materials and Sturtevant's use of *Drosophila* when the same breeding technique is used. Sturtevant took two of the new mutations that had been found to be unrelated to the sex chromosome. One of these, pink eyes, he tested with black and with vestigial, the two members of the second chromosome. In both cases, pink segregated in $9 : 3 : 3 : 1$ ratios. The other mutant, ebony body color, segregated in a $9 : 3 : 3 : 1$ ratio when tested with vestigial. "When pink-eyed flies were crossed to those with ebony body color the following result was obtained: gray red 3765; gray pink 1369; ebony red 1112; ebony pink O."[296] This F_2 result was approximately the ratio that would be encountered as a zygotic ratio for cases of complete repulsion (the $2 : 1 : 1$ ratio). "In the later generations from these crosses the writer was able to obtain a few ebony pink flies, and a pure stock of this combination was then made up. When these flies were used for 'coupling' experiments the result was: gray red 272; gray pink 12; ebony red 10; ebony pink 65." This zygotic ratio was similar to a partial coupling (in Bateson's system) with a gametic ratio of $15 : 1 : 1 : 15$. But using Morgan's hypothesis instead, Sturtevant reached a different conclusion. "That rather strong linkage exists . . . there can be no doubt."[296] The existence of "third," "second," and "X" chromosome groups of linked factors was significantly different from Bateson's interpretation because the reduplication hypothesis offered no explanation for this peculiar grouping of factors.

Bateson's reduplication hypothesis was lagging behind in productivity compared to the rush of publications coming from Morgan's laboratory during this same period. The chromosome theory enabled the Morgan group to use three or more factors at a time in the construction of a map. Fortunately for Morgan, the frequencies of crossing over in the "sex-limited" cases provided the standard for determining the frequency of crossing over for non-sex-linked genes. The double recessive forms obtained in the F_2 of such $F_1 \times F_1$ crosses were used to test cross the hybrid. This avoided the enormous difficulty that would be involved in $F_1 \times F_1$ crosses of the hybrids. Bateson, however, continued to use such $F_1 \times F_1$ crosses, and the progeny in the F_2 had to be checked empirically against tables of partial repulsion, zygotic series, for interpretation.

The reduplication theory was faced with other difficulties. Trow, in 1913, reported clear violations of the geometrical series for partial coupling and repulsion. "My own studies of *Senecio vulgaris* have . . . revealed the existence of the ratio $2 : 1 : 1 : 2$, and Baur . . . appears to have found the ratio $6 : 1 : 1 : 6$ in an *Antirrhinum* cross."[311] Trow attempted to predict a three-factor cross on the reduplication hypothesis. If a heterozygote, AaBbCc forms gametes from coupled parents, it would first undergo a segregation and primary reduplication of the factors A and B. The resulting cells would not be reduced for the factors C and c. The gametic ratio for the four cells, ABCc: AbCc: aBCc: abCc, would then have to undergo a secondary segregation and reduplication of the unreduced Cc. The primary reduplication geometric series would probably be different from the geometric ratios in the secondary reduplications (for example, the primary series might affect the coupling of AB in a $3 : 1 : 1 : 3$ ratio and the coupling of these four gametic types with C might be in a ratio of $7 : 1 : 1 : 7$). The combined effect of the two rounds of differential multiplications of cells would give eight gametic types with a complex ratio, the eight types being, ABC, ABc, AbC, Abc, aBC, aBc, abC, abc. Such a cumbersome gametic ratio for three factors would then have to be squared to give the F_2 zygotic ratio! It is obvious that doubly and triply recessive classes of zygotes would be swamped out by the enormous quantity of doubly and triply dominant classes among the progeny as a consequence of the $F_1 \times F_1$ mating. Only staggering counts in the F_2 would permit any hope of inferring the gametic ratios from the zygotic ratios.

Despite this limitation, Sturtevant attempted to test the validity of the predictive value of the reduplication hypothesis for the *Drosophila* data. Using the published papers of Gregory on *Primula* and Punnett's on the sweet pea, Sturtevant found that these data "are not entirely satisfactory because of the relatively small numbers involved and because in most cases the gametic ratios can be only approximately determined. This is due to the fact that most of the data concern F_2 counts from which gametic ratios cannot be calculated directly."[298] The two systems also did not agree in the predicted numbers in the zygotic classes (which would be the gametic ratios in the case of the X chromosome or in back-crossed experiments). Along with Trow's skepticism about the universality of the geometric fixed ratios proposed by Bateson and Punnett, Sturtevant had many instances, from Morgan's laboratory, of gametic ratios which departed from these geometric relations. Reduplication also forced the mechanism of segregation outside of the maturation divisions in meiosis.

All this was strong reason for rejecting the hypothesis. "We are forced to assume an enormously complex series of cell divisions, many of them differential, proceeding with mathematical regularity and precision, but in a manner for which direct observation furnishes no basis. It seems to me that it is not desirable to assume such a complex series of events unless we have extremely strong reasons for doing so. I can see no sound reason for adopting the reduplication hypothesis. It apparently rests on two discreditable hypotheses: somatic segregation, and the occurrence of members of the $3 : 1$, $7 : 1$, $15 : 1$, etc., series of gametic ratios in more cases than would be expected from a chance distribution."[298] As a consequence of this application of Occam's razor, "the chief advantage of the chromosome hypothesis of linkage . . . seems to be its simplicity. In addition it appeals to a known mechanism. . . . It explains everything that any of the forms of the reduplication hypothesis does and in addition offers a simple mechanical explanation for the fact that 'secondary series' are always smaller than Trow's special hypothesis calls for them to be. On the reduplication hypothesis this fact must merely be accepted, for, I think, it can not be explained."[298]

In this same year, 1914, Muller found "a gene for the fourth chromosome of Drosophila."[219] The cytological observations of Stevens and Wilson had reported four pairs of chromosomes in the female fruit fly. One of these chromosomes was very tiny. The numbers of genes in each chromosome, their map distances, and the remarkable association between the X and the "sex-limited" factors, all formed a strong case for the chromosome theory. "Granting then, the correspondence between size and number of chromosomes and of groups of genes, it is difficult to see why larger groups of genes should follow the distribution of the larger chromosomes unless we conceive the connection between the genes and the chromosomes to be that the genes are material particles actually lying in and forming a part of the chromosomes with which they go."[219] With the discovery that the mutant bent wings segregated independently of the factors on the X, second, and third, chromosomes, Muller could close the case for the theory of crossing over. "Thus the chief gap yet remaining in the series of genetic phenomena that form a parallel to the known cytological facts in Drosophila . . . has now been filled."[219]

For Bateson it would be another seven years before reduplication would be abandoned. In a letter sent to his wife after delivering a speech in Toronto, in 1921, Bateson wrote, "Only a line to say that last night's address went well. I had to announce my conversion on the main point, 'that chromosomes are definitely associated with the transferable characters' is how I express it. Much enthusiasm over this of course, but as a candid man, I don't see how any other view can now be maintained. . . ."[24]

Bateson, Castle, and Morgan: the Verdict of History

If we ask ourselves why Bateson and Castle were wrong, we can answer, of course, that experiments proved their hypotheses to be inadequate. But at the time of the formulation of their concepts, their hypotheses *did* cover the extant facts. Consequently, two rival theories, accounting for the same facts, must be judged aesthetically (e.g., for "simplicity") or by their predictive value. Of these two standards, the latter is always decisive. In the period before the critical tests can be made, which may involve years and depend upon new concepts and techniques, the participants and the spectators must make a choice. It is this period which foments polemics and provokes enmity; it establishes schools of workers committed by choice or by contagion to the patterning formulated by the chief protagonists. The sources of creativity then become stimulated by rivalry, loyalty, patriotism, or other motivations too diffuse to articulate. Creativity in the design of experiments extends or resolves the conflict. The resolution may favor the hypothesis that was originally the most complex; it may favor the hypothesis based initially on faulty and incomplete data; it may favor the work of an upstart whose own personality may have delayed the acceptance of his theories. Castle and Bateson admitted that they were wrong. Slowly, they have been eased out of prominence in textbooks, in reviews, and in the ultimate test of their utility for research—in the bibliographic references of the articles of subsequent generations. The victor slowly replaces his rivals in the course-work taught to a later generation of students, and as the participants fade into the past, a succession of victors and victories is presented as the basis for the development of the science. An illusion of progress, of discovery stimulating discovery, forms a pattern of history. Three rivals, equally admired in their own time, become trapped by the patterns and habits of their scholarship and research; a later generation judges: two are reduced to partial obscurity, the other elevated to near sanctity.

CHAPTER 8

The Presence and Absence Hypothesis

In the profusion of papers confirming the fundamental Mendelian segregation ratio of 3 : 1, numerous characters in a large number of organisms were investigated. Correns, in 1903, and Cuénot that same year, suggested that the recessive trait could be represented as the loss of the dominant characteristic. Using mice, Cuénot found that both black and gray mice were dominant to white mice. "The hybrids of the first generation are always, without exception, identical to the gray mouse . . . that is, the character, pigment, is dominant with respect to the character, absence of pigment."[93] So many instances of this phenomenon were recognized that Bateson, in 1906, considered it a general rule. "A generalization may . . . be attempted on different lines. It may be suggested that in the dominant type some element is *present* which is *absent* in the recessive type. The difficulty in applying such a generalization lies in the fact that not very rarely characters dominate which appear to us to be negative. . . . Consequently we are almost precluded from regarding dominance as merely due to the presence of a factor which is absent from the recessive form. Not impossibly we may have to regard such negative characters as due to the presence of some inhibiting influence, but in our present state of knowledge there is no certain warrant for such an interpretation."[17]

At the same Conference on Genetics where Bateson had announced his generalization as a presence and absence hypothesis, C. C. Hurst presented a case for presence and absence. "In cotyledon colours in peas, might not the character pairs be really presence and absence of yellow on a basis of green, rather than the contrasting yellow and green? Is it not possible that many of the so-called contrasting pairs of Mendelian characters are really compound, and that the true unit-characters are simply presence and absence?"[165] Hurst reported that ". . . out of a total of 44 pairs of Mendelian characters met with in my own experiments with plants and animals, no less than 41 . . . may be regarded as favourable to the hypothesis of presence and absence of unit-characters."[164] No clear-cut mechanism could be proposed, however. "With regard to the possible nature of the 'absence' factor, three distinct views present themselves: (1) there may be a concrete factor literally representing 'absence'; (2) the factor for 'absence' may represent simply 'presence' in a dormant or latent state; (3) there may be no factor at all, the presumed factor for 'absence' being simply nothing. . . . In the present state of knowledge it is difficult to say which of the three is the most reasonable. . . . The last view, that 'absence' is simply nothing, certainly appeals to the practical mind, and is perhaps, of the three, the one least open to objection."[164]

Criticisms of this theory took many forms. Castle objected to its applicability. "The assumption which underlies the explanation of color inheritance given by Cuénot, and adopted by Bateson, is that recessives *lack altogether* a certain factor necessary for the production of a dominant pigment; that albinos, for example, have one factor necessary for the production of pigment, but lack a second factor *altogether*. Now granting that such factors exist . . . it is perfectly certain that many albinos possess both of them. For albino guinea-pigs and Himalayan albino rabbits actually do form

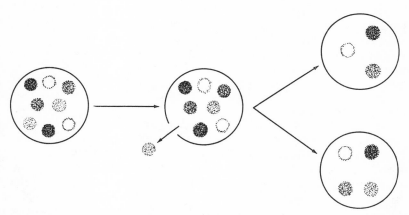

Figure 11. Bateson's Presence and Absence Theory. The four pairs of factors represented here segregate at each meiotic division. Occasionally a factor is lost. This leads to the formation of a cell that lacks the lost factor. In the gametes the allele of the light shaded factor is nothing—it is not present in the cell. The "recessive" condition consequently would be the loss of the "dominant" factor rather than an alternative (changed or mutated) form of the dominant factor.

hair pigments. There is nothing *altogether absent* from them which is a necessary factor in pigment production."[65] Castle, of course, had already committed himself to the variability of unit-characters and absence would not permit the allelic contaminations he envisioned. "The hypothesis of *absent factors* is inadequate to explain the observed facts, in at least a majority of known cases. By cross-breeding and selection we can alter the proportions of the different pigments in the coat *without eliminating* them."[65]

The possibility of dormant states, suggested in Hurst's analysis, was championed by C. B. Davenport. "I think it is clear that dominance in heredity appears when a stronger determiner meets a weaker determiner in the germ. The extreme case is that in which the strong determiner meets a determiner so weak as to be practically absent as when a red flower is crossed with a white. In such cases we have the clearest examples of Mendelian inheritance. But there is an entire gamut of cases where the opposed determiners are of varying relative potency. The phenomenon of determinance is seen in these cases also, but the Mendelian law in them is sometimes obscured and sometimes merely not applicable."[96]

The arbitrariness of Hurst's list of "present" unit-characters was criticized by G. H. Shull in 1909. "Yellow is described as present in the yellow pea and absent in the green pea. What is to hinder us from describing the green as present in the green pea and absent in the yellow one? Similarly in the contrast between tall and dwarf, one could perhaps say 'presence and absence of dwarfness on a tall basis' as appropriately as 'presence and absence of tallness on a dwarf basis.' "[276] Shull nevertheless adopted the presence and absence theory. "I will assume for the sake of discussion that the presence and absence hypothesis is correct and that the absence is real, having no internal unit to represent it. This assumption seems to me . . . to be simpler and more practical than the alternative idea that the alternative units are paired in the heterozygote, having a representative for absence as well as one for presence."[276] To avoid the use of *ad hoc* inhibitors to explain the cases of dominant absences, Shull advocated a chemical model of dominance. "Having arrived at the conclusion that all the Mendelian characters are dependent upon chemical relations, we may return to the question of dominance, and the relation between the two kinds of homozygote, and see to what extent the known facts may be interpreted in terms of chemical experience."[276] Shull argued that if one dominant "presence" factor performed a chemical reaction as efficiently as two dominant "presence" factors, then simple dominance would occur. If one dose of the factor gave only a partial reaction then a less intense or intermediate appearance of the heterozygote would result. In the case "in which the absence of a character is dominant over its presence . . . the character determined by A is latent in the heterozygote. . . . In the heterozygote where the chemical unit A (of whatever nature) occurs but once in each nucleus, no reaction becomes apparent, but in the pure bred forms bearing the unit A in double quantity, i.e., AA, the specific character (or reaction) produced by this unit appears. The heterozygote will then be indistinguishable from the negative homozygote. . . . In other words 'absence will be dominant over presence.' . . . For the sake of uniformity with the terminology adopted I may call this new kind of invisibility 'latency due to heterozygosis.' "[276]

In 1910 Morgan was opposed to the extension of the presence and absence hypothesis in the form of the "single chromosome: single character fallacy," but he had no objection to the general form of the theory as proposed by Bateson. "Characters that Mendelize are no longer allelomorphic to each other but each character has for its pair, the absence of that character. This is the presence and absence theory."[198]

Until Morgan advocated the existence of linkage in the form of the chromosome theory, all critics and proponents of the presence and absence hypothesis assumed that the unit-characters of an organism represented some sort of independent population of factors floating freely in a cell nucleus or in the cell itself. The simplest analogy would be colored glass beads in a fishbowl. If a bead were lost, its absence in a descendant cell would merely be reflected by one less bead in the bowl. But the subsequent character loss would be significant to the organism derived from such a cell. Coupling, as the name implied, would associate two of these beads in proportion to the forces that attracted or held them together. The beads, no longer being completely free-floating, would require some mechanism for their association. If the beads were faceted, then associations with other coupled beads could occur in varying degrees of strength. The facets by their regular shapes would determine a geometric series of coupling ratios. Bateson had long been fascinated by the subject of symmetry, and he pursued this aesthetically through an extensive collection of prints and drawings and also reflectively in some of his writings.

Bateson, however, could not and did not offer such a symmetrical model for his coupling ratios without a cytological basis for it. Repulsion could be incorporated into the fishbowl model with negative forces maintaining certain non-allelic unit factors apart. As long as such repulsion was complete, in the form of a spurious allelomorphism, this model or something comparable satisfied Bateson. When repulsion turned out to be partial, the fishbowl model fell apart. It would have required an ad hoc hypothesis of repulsive forces equivalent to attractive coupling forces. If the coupling was associated with a geometrical or symmetrical property of the factors in the nucleus, then the repulsive forces had to be of an entirely different basis but of the same magnitude! Furthermore, if coupling joined two unit-characters and if a single mechanism were invoked to account for both phenomena (as seemed probable), then the repulsion state would have joined a presence with its own absence. Clearly this was impossible on the presence and absence theory. How can a "something" be joined with "nothing" to give a specific ratio in repulsion crosses? Either the processes of association and dissociation—the implied fishbowl model—or the presence and absence hypothesis had to go. Bateson and Punnett rejected the fishbowl model and proposed reduplication, saving the presence and absence hypothesis as the more significant interpretation of the unit-characters in the cell.

With a mechanical model of the chromosome as a basis for "coupling and repulsion," Morgan and his students no longer had to assume the presence and absence of factors in the cellular or nuclear "void" that existed prior to 1910. But if presence and absence represented the process in which *parts* of the chromosomes were lost, parts corresponding to the factors for the presences, then chromosomes should be of unequal sizes in a heterozygote and pairing would be difficult to visualize. The normal factor in a chromosome could not pair with a gap! This consideration, however, was secondary to Morgan's discovery, in 1912, that a mutation in a white-eyed stock resulted in a fly with eosin colored eyes. Linkage tests showed that it was on the X chromosome and that it was near or at the place where white was situated.

Since 1900 it had been traditional for the discoverer of a new mutation to assume the existence of a corresponding allele for it—the normal factor. If eosin were a mutant which was unrelated to white, then in the cross of white to eosin, all the heterozygotes should be red-eyed (the heterozygote, WwEe, in which w is white, e is eosin, and the capital letters W and E are their normal alleles, would be normal for both factors involved in the eye coloration and hence the fly would appear red-eyed). When the cross was tried the F_1 females were a light eosin in eye color, intermediate between that of eosin and of white. Since eosin was recessive to its normal allele, this meant that the white stock *also* carried eosin. White, therefore, must have arisen as a *double mutant*, one change occurring at the eosin factor, the other nearby in a factor of unknown color, the two losses together removing all eye color from the fly. Eosin, one of these factors in white, was designated o and the unknown color factor was designated c. This made white homozygous ccoo; normal, red-eyed CCOO; eosin, CCoo; and the unknown color ccOO. Eosin had to arise then as a change from co to Co in one of the gametes from white (ccoo). The F_1 heterozygote from a cross of white to eosin was Ccoo. A test of this interpretation was immediately apparent. Red (CO) and white (co) in an F_1 (CcOo) individual, being coupled and heterozygous, should give four types of progeny in the F_2: red (CO), white (co), eosin (Co), and the unknown color factor (cO). The last two classes would be crossovers between C and O. Morgan and Bridges carried out a large-scale experiment in a search for the unknown factor and eosin from such a cross. The search was a failure. "We . . . assumed that the linkage between these two factors was well nigh absolute for the following reason: if a white male is mated to a red female there should appear in F_2, unless the linkage is complete, four types of males— red, white, *eosin*, and a *new* type which should be the single recessive, or simple white. We have records of approximately 150,000 flies from such a cross and its reciprocal and yet only two classes of males appeared—red and white."[213]

Sturtevant, in 1913, suggested an alternative to complete linkage as the explanation of this case. Complete linkage, preventing the isolation of the color factor, permitted only three of the four possible combinations for the two pairs of factors: red (CO), white (co) and eosin (Co). But if the factor for red mutated to white, why couldn't the factor for white, instead of being absent, have mutated to eosin? This would give two mutant alleles and one normal allele. "It will be seen that triple allelomorphs may be substituted for complete coupling as an explanation of any case where only three of the four combinations possible on the complete coupling scheme are known."[295] Morgan, agreeing with this new interpretation of multiple allelism, changed the symbolism for the white and eosin factors. White (originally W, then O for "absence," then co) became w; eosin (originally Co) became w^e; the normal red eye was then designated W (later it would become w^+). The confusion in symbolism arose from the various dual systems; the contrasting system used by Mendel (A and a as two alternative characters, one dominant and one recessive); the dual system used by Bateson in 1902 [D for some dominant trait, R for some recessive trait and DR or D(R) for the heterozygote, the recessive sometimes being placed in parentheses]; and Bateson's presence and absence designations in 1906 (A for a presence, a for its absence). With so many conventions to use, confusion abounded and Castle, in 1913, clarified the genetic symbolism by advocating the abolition of a dual system. In the future, all mutants that were recessive would be designated with a small letter and all new dominant mutants would receive capital letters. No capital letters would be used for the normal alleles of the recessive mutants (they would merely be "understood" to be there). The introduction of the superscript + for the small letters was later adopted as the convention for a normal allele. The individual which contained two mutant members of a multiple allelic series (the F_1 of the white x eosin cross) was no longer called a heterozygote. "The formula for the daughters of the above cross becomes then w w^e. We suggest calling this individual a 'white-eosin compound,' and the use of the term 'compound' as a name for such zygotes as are formed by the union of the mutant factors of a multiple allelomorphic system."[213]

Just before Sturtevant conceived of a multiple allelic interpretation of the white-eosin mutants, he presented the evidence for their "complete linkage" in a significant way. "C and O are placed at the same point because they are completely linked. Thousands of flies had been raised from the cross CO (red) by co (white) before it was known that there were two factors concerned. The discovery was finally made because of a mutation and not through any crossing over. It is obvious then, that unless coupling strength be variable, the same gametic ratio must be obtained whether, in connection with other allelomorphic pairs, one uses CO (red) as against co (white); Co (eosin) against co (white), or CO (red) against Co (eosin) (the cO combination is not known)."[297] This passage is revealing of the effect of both the presence-absence pattern of design and the pattern which would be dictated by the multiple allelic theory. The apparent complete linkage of white and eosin prevented the direct test for crossovers between them. This pair of alleles, however, would not have been chosen on the basis of a partial linkage theory because the terminology indicated the homozygosity of the eosin factor (the F_1 females being Ccoo). Recombine the terminology in any way possible, only eosin and white progeny could result from crossing over—the same as the parental forms that went into their composition. If multiple allelism were adopted, then on a presence and absence basis, white would be nothing and the "eosin-white compound" would only give eosin or white progeny. On the basis of a factor of a particulate or atomistic nature, recombination would separate particles and crossing over would occur to the left of the white-eosin alleles or to their right. Sturtevant suggested the one possibility that would have permitted the test, "that . . . coupling strength be variable" for the members of the series. There was no reason to invoke such a hypothesis, and the test was not made. If it had been taken seriously, crossovers between various white alleles would probably have been found in paired tests of that magnitude (150,000) and pseudoallelism would have been discovered as early as 1914! The consequences of such a discovery at that time would have profoundly affected the development of the gene theory.

The choice between complete linkage and multiple allelism was made by Sturtevant, "The question as to which of these two views is the more probable is closely bound up with the presence and absence hypothesis. On a strict application of this idea there is of course no possibility of more than two members of any given allelomorphic group."[295] On the complete linkage theory, the white-eyed fly ccoo underwent mutation of co to give a gamete Co which resulted in eosin. On the multiple allelic interpretation w mutated to w^e. On either hypothesis an "absent" factor (w or c) mutated to a "present" factor! This "reverse mutation" ruled out the total absence of the factor for white. The application of Occam's razor in this case wiped out two hypotheses in one swipe: presence and absence, which could not be maintained without serious modification, and complete linkage, which also required an ad hoc explanation for the intermediate coloration of the white-eosin compound (Ccoo was different in appearance from CCoo).

The factors were beginning to acquire properties that demanded a physical basis. Sturtevant was willing to give these factors a material reality, and to designate them as genes. "It seems very unlikely that protoplasm (chromatin?) is such a simple substance that the only possible change in a given unit (molecule?) involves the loss of that unit. On the other hand, if a slight change takes place in a chemically complex gene, is it necessary to suppose that its allelomorphic relations must be upset? That very slight changes in the constitution of a gene might easily affect its behaviour in ontogeny, will, I think, be readily granted."[295] The crucial evidence to equate factors with genes was not available in 1913. Morgan was convinced of the chromosome theory, but the conversion of the factorial hypothesis into the "classical" gene concept required another three years.

CHAPTER 9

From "Factor" to "Gene" Through Three Ph.D. Theses

In the five-year interval from 1912 to 1917, the factorial hypothesis met its challengers, led to new discoveries, and amassed complexity. Only two techniques were exploited—crossing over and cytology. The discoveries were so numerous, the findings so consistent, the implications so far-reaching, the versatility for research so subtle, that by 1917 Morgan was ready to accept the gene concept as a material reality and not as a mere "factorial hypothesis."

In 1912 Morgan was given a peculiar strain of wild-type flies. "At that time some females were producing two females to one male, and other females equal numbers of both sexes."[203] Morgan immediately suspected that the disturbed ratio had something to do with the X chromosome. "If sex is determined by a factor in the sex chromosomes it seemed probable that some change had occurred in this chromosome."[203] Morgan had the white-eyed stock to test this possibility; he mated individual wild-type females with white-eyed males. "Some of the F_1 females gave the 2 : 1 ratio. When these females were bred to white-eyed males again the following results were obtained: red ♀ 448, red ♂ 2, white ♀ 452, white ♂ 374."[203] The analysis of the data was brilliant. Morgan assumed from his model that the class of red males should be present; their absence therefore required the existence of a factor which prevented their appearance among the progeny. But if such a factor did exist, why were there any red males at all? "On the face of these returns it seemed likely that some lethal factor must be contained in the single sex chromosome of the males. . . . If the lethal factor contained in the chromosome in question should occasionally 'crossover' from the red factor, then a red-producing chromosome would result, which, if it went into a male, should give a normal male. . . . If, however, the lethal factor separates from the red-white factor (R-W) only once in 200 times it must be near that factor, on my hypothesis of the linear order of the factors in the chromosome."[203] Also, according to Morgan's hypothesis, the lethal factor, having been determined to be near white, should have given approximately the same frequency of crossing over with miniature wings as miniature would with white. "In brief, we predicted the ratio of long and miniature-winged males that are expected in the back cross, i.e., how many long winged males would escape the fatal dose. The prediction was verified."[203]

The ability to locate a mutant that could be inferred only from the absence of the progeny which contained it and to restrict the location of this factor to a known region of a particular chromosome in the gamete of a fruit fly was even more dramatic than the hypothesis of crossing over itself! It gave support to the superiority of Morgan's model over that of coupling or reduplication. The ratio of the F_2 progeny would have been inexplicable on any of Bateson's hypotheses. Ten years earlier Bateson had predicted the existence of lethal factors from Garrod's analysis of alkaptonuria and albinism, but his conception of genetic theory prevented him from making the discovery himself. Morgan, who, as late as 1910, would have scoffed at invoking a factor to account for *nothing*, was forced by his own model to generate new concepts that his critics would consider absurd and fanciful.

1913 is one of the greatest years in genetics. Sturtevant, extending Morgan's hypothesis, had proposed the first linkage map which would, in turn, eliminate reduplication as a competitor in the interpretation of coupling and repulsion. He also proposed a theory of multiple alleles which shattered the nearly universal acceptance of the presence and absence hypothesis. An even more dramatic discovery was to come from Morgan's laboratory, this time from C. B. Bridges. Both Bridges and Sturtevant were now working towards their Ph.D. degrees. Bridges was studying the white-eosin case with Morgan. When white-eyed females were crossed to red-eyed males the F_1 daughters were red-eyed and the F_1 sons were white-eyed. This type of inheritance was called "criss-cross" inheritance by the Morgan laboratory because the sons were like their mothers and the daughters were like their fathers. Criss-cross inheritance was also found to hold true for eosin. However, Bridges discovered imperfections in some of his sex-linked cases. "From time to time in an F_1 where sex linked characters were concerned, females or males have arisen which I was unable to explain on any current Mendelian hypothesis. Some of these I bred, and the offspring were as hard to explain as the original exceptions."[49] The first case, a female, violated criss-cross inheritance. "In the exceptional case that I have found, where the mating was like that just described, about 5 percent of the daughters are like the mother and 5 percent of the sons are like the father."[49] The critical cross involved a P_1 female with a third chromosome factor, pink eyes, mated with males that had white eyes and miniature wings on the X chromosome and black body color on the second chromosome. Among some 550 progeny in the F_1 there were 3 white miniature males. "Fortunately I bred one of these males to virgin sisters, which were heterozygous for white, miniature, black, and pink. The results showed conclusively that the male had really come from the cross and was not the result of any contamination, in that it gave in the next generation all expected classes from the cross of an F_1 female by a white miniature male heterozygous for black and pink, that is, there appeared blacks and pinks, males and females, whites and miniatures, in all the combinations and permutations expected."[49] Bridges interpreted his exceptional cases as cytological abnormalities which had occurred during meiosis. Ordinarily the paired X chromosomes in the female separate or disjoin from one another; rarely the two will fail to disjoin and the two X's will go to one daughter cell and no X will go to the other cell. An egg bearing two X chromosomes, encountering the male determining sperm (which would contain no X, on the basis of Steven's 1908 observation) would produce a female. An egg with no X chromosomes, when fertilized by a female determining sperm (i.e., bearing the paternal X chromosome) would produce a male which, in violation of criss-cross inheritance, looked like the father. This failure of chromosomes to disjoin Bridges called "non-disjunction."

In 1914 the remaining peculiarities in the tests of his exceptions forced Bridges to review the cytology of *Drosophila* and he made a startling discovery. The male was not XO but had two "accessory" chromosomes—the solitary X and a morphologically distinct, hook-shaped chromosome which he designated as a Y. The Y chromosome was anomalous. It did not appear to have any normal alleles for the various recessive mutants known for the X chromosome. Surprisingly, it was not even needed for the viability of the flies! Many of the exceptional non-disjunctional males were lacking a Y chromosome. Also, it was not the male determining chromosome because the non-disjunctional females contained both of the maternal X chromosomes as well as the paternal Y. These females were perfectly fertile. The only attribute of the Y, other than its existence in the normal male, that Bridges could find was its necessity for male fertility. The "primary non-disjunctional" males which he and Morgan had encountered in the eosin and white crosses were sterile. All these cytological findings corresponded to the genetic findings, and Bridges was delighted to offer a Ph.D. thesis giving "direct proof through non-disjunction that the sex-linked genes of *Drosophila* are borne by the X-chromosome."[50] Furthermore, Bridges's thesis was published in 1916 as the first paper in the new journal *Genetics*.

Sturtevant's Ph.D. thesis was also a major contribution to the chromosome theory. All the available mutants in *Drosophila*, numbering about two dozen in 1914, were mapped onto their appropriate chromosomes, each map constituting a linkage group, any one of whose members would segregate independently from the factors on any of the other linkage groups. Muller's finding of the fourth chromosome factor, bent wings, completed the linkage groups. The distribution of the mutants, however, was more exciting than the mere fact that they could be localized to specific sites or "loci"; the distribution gave an insight into the gene concept itself. "Although there is little that we can say as to the nature of Mendelian genes, we do know that they are not 'determinants' in the Weismannian sense. This is well shown by the following case. The difference between normal red eyes and colorless (white) ones in Drosophila is due to a difference in a single gene. Yet red is a very complex color, requiring the interaction of at least five (and probably of very many more) different genes for its production. And these genes are quite independent, each chromosome bearing some of them. Moreover, eye-color is indirectly dependent upon a large number of other genes, such as those on which the life of the fly depends. We can then, in no sense identify a given gene with the red color of the eye, even though there is a single gene differentiating it from the colorless eye. So it is for all characters—as Wilson (1912) has put it '. . . the entire germinal complex is directly or indirectly involved in the production of every character.' All that we mean when we speak of a gene for pink eyes is, a gene which differentiates a pink eyed fly from a normal one—not a gene which produces pink eyes per se, for the character pink eyes is dependent upon the action of many other genes."[299] This conception of the gene could only have gained acceptance through the mapping of the genes. In reduplication or coupling and repulsion no unit-character could be given such a qualification. Without mapping to reveal the scattered locations of the mutants at their specific sites, the individual character could have been attributed to one aggregate ("partially" coupled or repulsed) or to one common particulate type.

In 1916 Muller published his Ph.D. thesis; the "mechanism of crossing over" acquired more complexity. When the first maps were made on the X chromosome, it was thought that the maximum percent of crossing over obtainable (fifty percent of crossovers among the progeny) would correspond to fifty map units on the map. The actual length of the map, however, for the second and third chromosomes turned out to be more than fifty map units. Also the length of two genes far apart when measured by a direct test, was always smaller than the length obtained by summing up all the intervening distances defined by the other known genes between the two that were far apart. To account for this, Muller suggested that the great distance between two coupled genes might experience two crossovers so that the genes outside the boundaries of these crossovers would remain coupled. Such "double crossovers" would give the illusion that no crossing over had taken place unless a third gene was introduced between the two points of crossing over. In that case the middle gene would merely be shifted out of its association with the other two genes (proving that a double crossover had actually occurred). Assuming that all crossovers could occur at random along the length of a chromosome, Muller calculated the probability of double crossovers for two adjacent segments and compared these to the actual numbers of double crossovers obtained. The results were strikingly different from what was predicted! For small distances no double crossovers could be detected at all; for intermediate distances (ten to thirty map units) a sizable number of double crossovers were present, and only for the larger distances would tests show a close correspondence between the predicted and actual double crossover classes of progeny. "In a sense then, the occurrence of one crossing-over interferes with the coincident occurrence of another crossing-over in the same pair of chromosomes, and I have accordingly termed this phenomenon 'interference.' The amount of interference is determined by comparing the actual percent of double crossovers with the percent expected if crossing over were independent, i.e., if they had a purely chance distribution with reference to each other. Now the percent which would occur on the latter expectation has . . . been given . . . as percent AB × percent BC. If then, the observed percent of double crossovers were divided by percent AB × percent BC, we would obtain a fraction showing what proportion of the coincidences which would have happened on pure chance really took place. . . . The ratio is itself best expressed in per cent, and it may be called the relative coincidence, or simply 'coincidence.' "[220] Muller attributed the interference in double crossing over to a release of tension of the chromosomal twists whenever a crossover occurred. If the tension were released, its immediate effect would be the inhibition of additional breaks nearby, but the further away other twists were, the less would they be affected by the tension changes of the other crossover, and their probability of crossing over would thus remain relatively unaffected.

From the more refined mapping of the several dozen mutants available to him, Muller also recognized that crossing over could give some rough approximation of the number of genes in Drosophila. "Presumably the factors are set very close together in the line, judging by the fact that mutations in new 'loci' (positions in the line) are still as numerous as ever, and that, if the whole chromosome is packed with factors as close together as, judging by their linkage relations, they seem to be at certain places in it, it must contain at the very least 200 factors."[220] This estimate was too low because the sizes of the maps were limited by the available mutants at the tips of the maps. As new mutants were found to the left or right of the linkage maps, the maps were augmented and the number of factors, estimated by Muller's method, would also increase proportionately.

New cytological findings concerning the chromosomes made Morgan's linearity of the factors seem likely. Previously cytologists had thought the rod-like chromosomes seen during the height of cell division (metaphase and anaphase) fragmented at the end of cell division and dissolved into a diffuse "chromatin network" which somehow mysteriously reaggregated

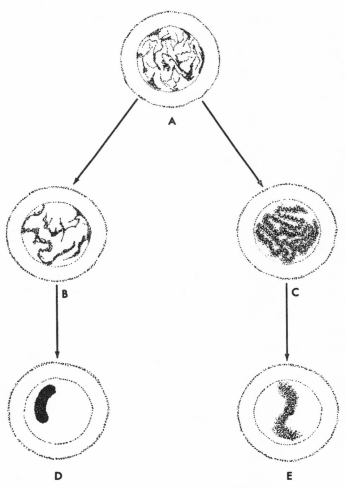

Figure 12. The Integrity and Continuity of the Chromosome. Until late in the decade following 1910, it was commonly believed that the chromatin (A) was an irregular network which coalesced (B) with the onset of cell division. The coalescence eventually formed the chromosomes (D). In contrast to this view, Bonnevie and Vejdovsky asserted that the chromatin was actually a long unwound thread which became visible (C) through coiling. The compact coiling in cell division gave the chromosome (E) its morphology. This cycle of coiling and uncoiling preserved the integrity and continuity of the chromosome from one cell generation to the next.

into their solid forms during the next cell division. But "the observations of Vejdovsky and others, taken in connection with the genetic results from Drosophila, render it practically certain that the factors are really disposed in an extremely fine, long thread or 'chromonema,' which, during the metaphase and anaphase of mitosis, is coiled up very closely in more or less spiral fashion . . . to form the thick dense chromosomes, but which, in the resting period and during the early stages of synapsis, becomes, to some extent at least, uncoiled and drawn out again."[220]

One additional observation reaffirmed Muller's distaste for Castle's theory of allelic contamination. "Incidently, the results demonstrate another point, lying in a somewhat different field of genetics. By following the method of keeping stocks constantly in heterozygous condition, twenty-two factors have been continually outcrossed, in each successive generation, to their allelomorphs. Yet after about seventy-five generations of outcrossing, these characters do not show the slightest contamination. The experiment therefore forms an extensive test and verification of the 'purity of Mendelian segregation.' "[220]

Another unexpected case was found in 1917 by Bridges. A stock which had the dominant mutation, Bar eyes, produced a fly with round eyes and at the same time appeared to have lost its normal allele for forked bristles. At first it was thought that a piece of the chromosome was lost, as if a toothpick were snapped into two pieces. Tests showed, however, that the loss was not that extensive but that it did include several genes near the Bar locus. Bridges called this multigenic loss a "deficiency." "This case of deficiency therefore constitutes the first valid evidence upon the question of 'presence and absence.' And it is significant to notice that the occurrence of the deficiency of a considerable section of genes has not brought to light any visible mutative changes in the way of dominants, contrary to what might well be expected on the presence and absence hypothesis. This is the more significant when it is recalled that the deficient region included the locus for bar—a known dominant mutation. According to the presence and absence hypothesis the original appearance of the dominant bar character was due to the loss from the chromosome of an inhibitor, thereby allowing the normal narrowing effect of the remaining complex to assert itself. . . . If, however, the appearance of the bar character were due to the creation of a new presence, then of course the loss of this presence by deficiency should restore the original condition: but that advocates of 'presence and absence' have little liking for this type of explanation of the origin of a dominant is evident from the lengths they go in some recent expositions to avoid the vexed questions of the origin of presences."[52] Bridges studied the results of crossover tests using his deficiency. "Not only did the deficiency mutation affect a section of adjacent genes but it also removed the crossing over from a definite section of chromosome. Now when the section from which the crossing over has been removed is compared upon the map with the section in which the genes are affected the two are seen to be identical. For a definite section of genes an identity has been established between the map and an actual distribution of genes."[52]

Also in 1917 Muller presented a sensational analysis which served four purposes. It unraveled the mystery of an "inconstant" factor, beaded wings; it proved the existence of the "residual heredity" which caused character variation; it gave an explanation for de Vries's theory of mutations, and it introduced a new concept for "perpetual hybrids." Beaded was a dominant mutation causing scalloped excisions of the edges of the wings. The effect was variable and it could be modified by selection. When Beaded was crossed to normal flies, no more than half the progeny would show Beaded. After several years of selection a line of Beaded was isolated which showed almost 100 percent Beaded progeny. A few non-Beaded progeny would appear at irregular intervals among the later generations. Muller argued that Beaded was a dominant visible mutation that was lethal as a homozygote; he claimed its variability was due to other factors which he actually mapped in the various chromosomes of *Drosophila*; he argued that the "stable" line of Beaded was due to another lethal in the homologous chromosome which was like Morgan's 1912 lethal—a simple recessive. The consequence of this was a state of "balanced lethality" which killed either of the homozygous progeny but permitted the viability of the heterozygote. The "escapers" with normal wings were crossovers between the two lethals (like the red males that "escaped" death by crossing over between white and Morgan's lethal). "This remarkable genetic situation, wherein both types of homozygotes are prevented from appearing by the action of lethal factors in opposite chromosomes may be termed a condition of 'balanced lethal factors.' "[221] If de Vries's *Oenothera* plants were similarly balanced lethals, they would produce crossovers whose newly expressed recessive characteristics, long kept heterozygous, would manifest new character changes of a far more extensive range than the simple mutations usually encountered in *Drosophila*. Furthermore, de Vries's "mutations" would not be mutations! They would be crossovers!

Muller was also able to infer from the number of recessive lethals that were found that they were exceptionally more frequent than other recessive changes. "Following up, now, our original inference regarding the high frequency of lethals among the recessive mutants, it should further be pointed out that since recessive mutants as a class are much more numerous than dominant mutants, recessive lethals also should arise much oftener than dominant ones."[221] The significance of the Beaded case extended to the scientific attitude itself. "The beaded case illustrates to great advantage the danger of confusing characters with gens* and of drawing radical conclusions concerning the behavior of gens on the basis of uncritical experi-

---

* Muller (or the editor) used gen and gens for gene and genes. Shull pointed out the accuracy of using the final e for proper pronunciation. "Purists" had objected to the final s on genes because the German word *gene* was plural for *gen*. The literature between 1910 and 1920 makes frequent use of the archaic spelling of this term.

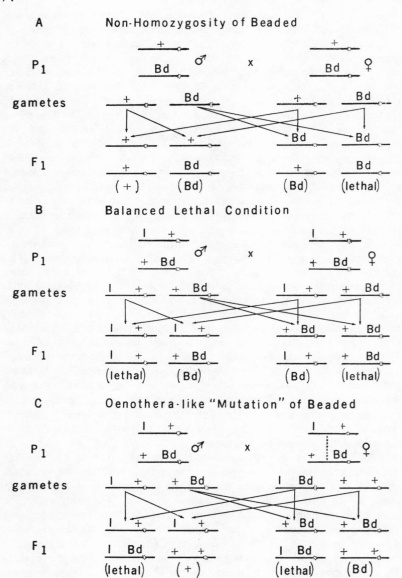

Figure 13. Beaded Wings and Balanced Lethals. In A, Beaded flies cannot produce homozygotes because the homozygote dies. An apparent homozygosity occurred in B when an independent lethal, without visible effect, arose in a homologous chromosome. The homozygous lethal (1) and the homozygous Beaded both die. This leaves only the heterozygous Beaded progeny. This balanced lethal condition, Muller claimed, was analogous to the genetic condition of the *Oenothera* species. Apparent "mutations" (in C)

ments. The work of the first four years upon the inheritance of beaded wings gave evidence which would to many have appeared most elaborate and convincing, that the hereditary material in this case was fluctuating and miscible, consisting of vague and plastic 'tendencies,' rather than definite physical particles." And in a caustic appraisal that Bateson might have wielded at Weldon, Muller charged that "here analysis has made the Mendelian machinery at work evident. Unwelcome as these conclusions may sound to obscurantists and to those in general who have an antipathy for exact modes of procedure, the necessity for such refined methods should be obvious."[221]

With this armament of information from his students and from his own participation in the analysis, Morgan was now a strong advocate of the "physical basis of heredity." In 1916 he gave the factor (conceptually translated into a point or locus on a map) some potential for change. "On a priori grounds there is no reason why several mutative changes might not take place in the same locus of a chromosome. If we think of a chromosome as made up of a chain of chemical particles, there may be a number of possible recombinations or rearrangements within each particle. Any change might make a difference in the end-product of the activity of the cell, and give rise to a new mutant type. It is only when one arbitrarily supposes that the only possible change in a factor is its loss that any serious difficulty arises in the interpretation of multiple allelomorphs."[214] Morgan was ready for all critics; in 1917 he took them all on in "The Theory of the Gene." "It has been said, for instance, that the factorial interpretation is not physiological but only 'static,' whereas all really scientific explanations are 'dynamic.' It has been said that since the hypothesis does not deal with known chemical substances, it has no future before it, that it is merely a kind of symbolism. It has been said that it is not a real scientific hypothesis for it merely restates its facts as factors,—then by juggling with numbers pretends that it has explained something. It has been said that the organism is a Whole and that to treat it as made up of little pieces is to miss the entire problem of 'organization.' It has been seriously argued that Mendelian phenomena are 'unnatural,' and that they have nothing to do with the normal process of heredity in evolution as exhibited by the bones of defunct mammals. It has been said that the hypothesis rests on discontinuous variation of characters, which does not exist. It is objected that the hypothesis assumes that genetic factors are fixed and stable in the same sense that atoms are stable, and that even a slight familiarity with living things shows that no such hard and fast lines exist in the organic world."[207]

To this set of "indictments" Morgan marshaled the entire weight of

---

would correspond to crossovers in a balanced lethal condition. The homozygosity of accumulated visible mutations brought about by crossing over would give the rare crossover many new features distinguishing it from its parent. In principle, Muller was correct, but the balanced lethal condition in Oenothera was even more complex than this crossover model.

his students' findings, and of his own thinking on the numerous factors which were generated from *Drosophila* in a steady profusion over the preceding seven years. The factorial hypothesis met every one of the challenges. "The germ plasm must, therefore, be made up of independent elements of some kind. It is these elements that we call genetic factors or more briefly genes."[207]

CHAPTER 10

The "Classical" Gene

Whenever a new scientific concept comes into prominence, it sends shock waves of surprise to the scholars contributing to that field. For some the conceptual patterns of old are so shattered that they leave the company of productive scholars and content themselves in other scientific pursuits or use their other talents to provide new outlets for their careers. For others the first shock waves generate disbelief, if not resentment, and the theory is ridiculed; it is made to apply to all situations familiar to the old patterns or to new alternative theories, and the first evidence of inconsistency is immediately rushed to press to contradict the threatening theory. For those who have no fixed patterns, and for those weary of old and unproductive patterns, the new concept becomes a rallying cause; its champions raise the concept to classical status and fill out the voids and implications which the concept, in its original form, was too untried to predict.

The chromosome theory, flourishing through Morgan's school, was attacked in 1917 by Richard Goldschmidt.[142] Goldschmidt could not understand how a linear linkage could be proposed for genes when they dissociated between cell divisions and were re-established as chromosomes by some sort of cohesive force ("Kraft") in the next cell division. It was just as reasonable to attribute the properties of linkage to this force as to the physical distance between linked genes. In Goldschmidt's scheme the chromosomes never crossed over. They served as a housing for the genes, which would leave their sites at the completion of cell division and enter into cell metabolism. On their return the sites along the chromosomes would produce forces whose specificity corresponded to the genes and all the genes would return to their proper sites. If there were a recessive allele

77

of a gene, its force would be comparable to that of its normal allele but sufficiently different so that occasionally it would occupy the site normally held by the normal allele. Such a mistaken occupation of a site would give the impression that a crossover had occurred with respect to the other genes in that chromosome. The frequency of "mistaken occupancy" would be proportional to the "variable force" exerted by the recessive gene. The less two alleles resembled one another in their force patterns, the "closer" their linkage with other genes on a map, since mistaken occupancy would be slight. The more the allelic forces overlapped, the further their apparent linkage with other genes on a map, because mistaken occupancy would be frequent. The linkage map would then only be a convenient representation of the relative forces of genes and would have no relation to the actual positions of the genes in their chromosomal housing.

Sturtevant, in sharp reply, repudiated this interpretation "The argument is based on the 'von jedermann anerkannten Voraussetzungen der chromosomenlehre (assumptions concerning chromosome principles acknowledged by anyone).' Among these Voraussetzungen Goldschmidt includes the idea that the chromosomes lose their structure during the resting stages, so that it is necessary that the particles be reassembled later to form the chromosomes seen at mitosis. . . . This idea forms the basis of Goldschmidt's whole argument, for it is assumed that the same mysterious 'Kraft' is responsible for the rebuilding of the chromosomes and for crossing over. . . ."[300] The fact that double crossovers occur was evidence against this, Sturtevant argued, because the middle section between the two pairs of genes crossing over would exchange places with the rest of the chromosome (as the intervening genes in that segment had shown in several cases from the Morgan laboratory). If one chromosome was abcdef and its homologue ABCDEF, then Goldschmidt's cases of double crossovers should be aBcdEf and Sturtevant's cases would be aBCDEf. "Goldschmidt declines to discuss the doublecrossover data further 'weil wir glauben dass die Sturtevant'schen Vergleichzahlen auf grund falschen Formel berechnet sind, und sodann weil es . . . gar nicht unsere Absicht ist, das hier benutzte Schema an Stelle des Morgan'schen setzen zu wollen (because we believe that Sturtevant's calculations are specified on the basis of a false concept, and therefore it is not our intention to employ that model in place of Morgan's).' Under the circumstances it seems natural to expect some cogent reasons to be given for this denial. No explanation of linkage can have any claim to serious consideration unless it accounts for these facts."[300]

Bridges pointed out an "intrinsic difficulty for the variable force hypothesis of crossing over"[51] which destroyed its usefulness. The alleles G and g for specific sites F_G and F_g ordinarily are housed properly: F_G-G and F_g-g. If one percent crossing over occurs (or a slight overlapping of the variable forces) then the switched housing relations would be F_G-g and F_g-G. But then in the next generation there should be ninety-nine percent crossing over since F_G and G will house properly once again and so will F_g

and g. Goldschmidt's model had no mechanism to prevent such a reversal of crossing over back to the original coupled condition. The "classical" concept proved the stronger in this case because Goldschmidt's model, like Bateson's reduplication, was limited to a few simple situations and lacked the adaptive features of "linear linkage."

Castle, whose controversial "contamination" theory persisted for almost sixteen years, was soon embroiled in another conflict with Morgan's school. "That the arrangement of the genes within a linkage system is strictly linear seems for a variety of reasons doubtful. It is doubtful, for example whether an elaborate organic molecule ever has a simple string-like form."[77] Furthermore the distance relations were not in agreement with the published data. "In reality it has been found that the distances experimentally determined between genes remote from each other are in general less than the distances calculated by summation of supposedly intermediate distances. . . . To account for this discrepancy Morgan has adopted certain subsidiary hypotheses, of 'interference,' 'double crossing over,' etc. . . ."[77] But "if the arrangement of the genes is not linear, what then is its character? This query led me to attempt graphic representation of the relationships . . . but finding

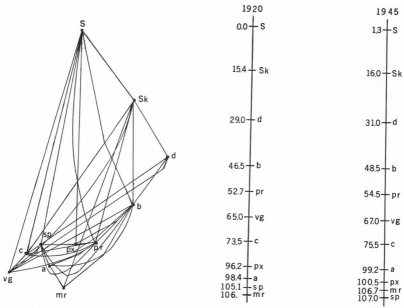

Figure 14. Castle's "Rat Trap" Model. Morgan and his students aligned their genes linearly on a map. Castle used their data to construct a three-dimensional model based on actual recombination frequencies. The difference between the two maps reflects the use of double crossing over (interference) in the linear map and the refusal to accept this procedure by Castle. The 1920 map has not changed appreciably since it was last revised (1945).

this method unsatisfactory, I resorted to reconstruction in three dimensions, which has proved very satisfactory."[77] Castle's reconstruction consisted of wire lengths attached to eye screws. Each length was scaled to the actual reported crossover frequency between a pair of genes. The genes themselves were represented by the eye screws. The finished model resembled a three-dimensional web (or more derisively, among the members of Morgan's laboratory, it was nicknamed the "rat trap" theory). "What, it might be asked, does this reconstruction signify? Does it show the actual shape of the chromosome, or at any rate of that part of it in which the observed genetic variations lie? Or is it only a symbolical representation of molecular forces? These questions we cannot at present answer."[77]

Castle nevertheless conceived of the chromosome as a single molecule, three-dimensional and rod-like in shape. "If the genes are not arranged in a single linear chain, the chiasmatype theory will need to be reëxamined. Such a pure mechanical theory seems inadequate to account for the interchange of equivalent parts between twin organic molecules, such as the duplex linkage systems of a germ-cell at the reduction division must be. Twin molecules are now closely approximated and parts of one may leave their formed connections and acquire new connections with the corresponding part of the other twin. . . . It is like the replacement of one chemical radicle with another within a complex organic molecule and it seems highly probable that such is its real nature."[77] The theory of crossing over, in contrast to this model, was inadequate, according to Castle to account for the discrepancies in map distances and had to be "saved" by invoking double crossovers. Double crossing over "is accordingly a secondary hypothesis of linear arrangement, but it can not be cited as proof of that hypothesis, which must stand or fall on its own merits."[77]

In an article "Are the Factors of Heredity Arranged in a Line" conspicuously authored by "Dr. H. J. Muller," Castle's concept was taken to task. Double crossing over, far from being a secondary hypothesis, was required by the theory of crossing over; otherwise a special mechanism would have to be postulated to prevent more than one break! Castle's assumption that no double crossing over occurred was then a secondary hypothesis. "For, once the occurrence of single breaks in a chromosome is admitted—a point agreed upon by both sides—it is just as arbitrary to deny the possibility of double breaks as to assert their existence."[223] If the chromosome were a three-dimensional rod with the genes distributed inside it like peppercorns in a sausage, then the linearity of the factors would be the reverse of what linkage actually encountered. "It is evident . . . that more remote factors . . . are likely to be arranged more nearly in a straight line than factors nearer together. . . ."[223] Having examined all Castle's objections and finding counter-arguments for each, Muller concluded, "The idea that the genes are bound together in a line, in order of their linkage, by material, solid connections thus remains as the only interpretation which fits the genetic findings. In view of the additional fact that the chromo-

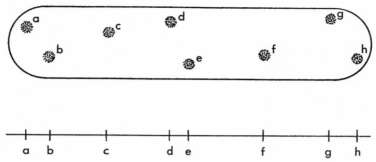

Figure 15. Muller's Criticism of "the Rat Trap Model." If genes are distributed in a three-dimensional rod-shaped chromosome, the non-additivity of close genes (a, b, c) is more extreme than the non-additivity of the genes farthest apart (a, e, h). In the linear scheme of the Drosophila group, the linearity of close genes (a, b, c) is obtained by experiment (ab + bc = ac) but non-additivity increases with genes which are farther apart (ad + dh is more than ah). The data are consistent with linearity and double crossing over, but they contradict the predictions of the rat trap model.

somes—themselves known to be specifically linked to the factor groups—can, at certain stages of their history, be seen to have the linear structure required, it would indeed be rash to adopt a different theory, without the most cogent evidence of a startlingly new character."[223]

The controversy was short-lived. The following year, 1920, new data from Gowen's analysis of third chromosome linkage were placed into Castle's wired model. "I next examined the data given by Gowen for the linkage relations of eight genes of the 'third chromosome' of Drosophila. On constructing a model to show the spatial relations of the genes, without assuming the arrangement to be linear, and without making any allowance for double or triple crossing-over, I obtained a figure strikingly similar to the model of the sex-linked group of genes. The eight genes fell into a curved band lying nearly in one plane, the model being quite flat as viewed edgewise. . . . Obviously the arrangement approaches the linear, but it is curvilinear rather than rectilinear as shown in the chromosome maps. But if we grant that the arrangement is in any sense linear, then it must be granted also that double and triple crossing-over are likely to occur. . . . But if double crossing-over occurs oftenest when long distances are involved, it follows that the long distances will have been apparently shortened. . . . I conclude that the model supports the linear hypothesis, if it be supposed that the longer distances have been shortened by double crossing-over, and that map distances in such cases should exceed observed cross-over percentages."[78]

At the time of this controversy over linkage models, Muller was developing a new tool for the study of the gene concept. In mutation, Muller hoped to characterize the gene. How stable is the gene? In what ways could it mutate? Would multiple alleles always form a quantitative series resembling partial losses? Was mutation a spectrum of plus and minus changes around a norm for each gene? Was mutation limited to one stage of the life cycle? What were the frequencies of mutation in different organisms? What type of defect produced a mutation? The answers to these questions were not known in 1919. "There is, to be sure, enough work to show that the real mutations are 'rare'—whatever that term may mean, but, so far as an approximate quantitative determination of the rate of factor change is concerned, it is not possible, from the published work, to determine even its general order of magnitude."[250] With the participation of Edgar Altenburg, who was also a product of the Columbia University genetics group, Muller introduced a new tool for analyzing the mutation problem. "There is reason to believe that by far the commonest type of mutation is that which gives rise to a lethal factor—which kills the organism containing it— and such lethal factors . . . in the X chromosome of a female, would be revealed, in this case, by the fact that half the male offspring, receiving it, would die before hatching."[246]

Using wild-type strains from which pre-existing lethals were screened out, Muller and Altenburg found a far higher frequency than they had expected: 13 of 385 females gave progeny in the 2 : 1 ratio characteristic of lethals. A second series, in which the wild strain of females was mated to males carrying several sex-linked recessive genes, produced 20 lethals among 1062 progeny. "Enough work has been done on them thus far to know that they occurred in at least ten different loci scattered along the X chromosome—but this is a bare minimum. Four of the lethals (perhaps five) are more strictly speaking 'semi-lethals,' as they occasionally allow the male possessing them to live (and then produce some curious morphological effects in him) but lethal mutations are so much more frequent than the type of visible character variation ordinarily dealt with, that none of the latter were observed in the whole experiment."[246]

One of the rarer qualities of the human mind is its capacity to extend a concept in directions that would not be suspected on superficial examination. In its most unusual form that rare quality surpasses even this achievement; it constructs patterns unseen by the most patient inspection; its novelty verges on the fanciful; it gives its possessor a prophetic vision. That quality—call it genius—was expressed in the next four years in four papers on the gene. Through these papers Muller made the gene a "classical" concept: it acquired a timeless characteristic. Extended into the past it formed the basis of evolution and of the origin of life itself; in the immediacy of the present it gave "the gene as the basis of life," rejecting "protoplasm" as a vague and ill-defined concept; for the future it generated the direction of future research, revealing new pioneering fields and the techniques that would be needed for their exploration; for the future too, it gave an insight into the directions in which evolution could go. Like Pandora's box, the gene through the mind of the genius unleashes a horde of implications which are both awesome and prophetic.

In 1919 the problem, in its simple form, was a matter of determining the mutation rates of genes. Its extension, in Muller's analysis, went beyond the collection and reporting of data. "If we accept the one in fifty-three ratio as representing the average frequency for the X chromosome, and if, as there is reason to believe, mutation occurs at the same rate in the other chromosomes as in the X chromosome, then, since the X's form about one-fourth of the entire chromosome mass, we may figure that about one fly in every thirteen has a new lethal mutation in some chromosome or other. It is evident that, at this rate, without natural selection to weed out the 'unfit,' the race would soon become filled with lethal factors."[246] Departing from the fruit fly's plight, Muller turned to a new organism, "The rate of change for the X chromosome in *Drosophila* is thus about one detectable mutation in four years. This immediately shows us that *Drosophila* must have a different rate from some other organisms—man for example—for if the X chromosome of man mutated at anything like a similar rate, all the X chromosomes in a female would contain several lethal factors by the time she was ready to reproduce, and none of her sons would be viable."[246] Earlier, Muller had estimated some 200 factors in the X chromosome of *Drosophila* on the basis of the maps available in 1915; the estimate in 1919 had risen to 500. "Taking this undoubtedly much too low minimum figure, it is easy to see that, if 500 factors show only one mutation in four years, each individual factor must on the average show a change in its composition only once in 2,000 years (yet this is in the mutable Drosophila)."[246]

In 1920 the number of multiple alleles in the white-eye series had grown to ten. Their time of origin was diverse. A new white, indistinguishable in color from Morgan's, arose as a mosaic individual. The male possessing it had one red and one white eye. Its progeny indicated that all the germ cells in its testes bore X chromosomes with the factor for the new white. This mutant had to arise during the very earliest cleavage stages of the zygote. The allele *ecru*, in contrast, arose as a solitary mutant male among its sibs. Its origin was late in oögenesis, perhaps in the ovum itself. The allele *ivory*, arose as a "cluster" of males descending from one female, most of whose X chromosomes were normal. Such a "cluster" must have arisen in early oögenesis, while the oögonial cells were still proliferating mitotically. Most geneticists believe, however, that the solitary condition, being the most frequently encountered type, could be interpreted as arising from a highly mutable stage at or near the end of meiosis. But, "there is no evidence that mutations are more likely to occur in gametes or germ cells near the period of maturation than in cells at any other stage in the life cycle. The peculiarities of cell lineage would, however, provide a greater chance for the appearance of single mutant individuals than for cases of twin or multiple mutants; since there are more cells existing during the later stages of the germ-cycle, there is a correspondingly greater chance for mutations to occur there in one cell or another, even though mutations may occur equally readily in cells at any stage."[224]

The spectrum of color changes made unlikely any simple model of mutation; "Of the ten certain variations of W, it is first noticeable that all were 'minus' variations—i.e., lighter than normal, in spite of the fact that 'plus' variations in eye color can and do occur in other loci. This immediately removes the curve of variation from the ordinary symmetrical bell-shaped class and would make it an extreme case in the class of 'skew-curves.' The second outstanding feature is that three of the ten mutants were of the most extreme possible types, namely, white . . . while four were very nearly white. . . . If anything, then, there is a piling up of the curve of variation near its extreme—a phenomenon diametrically opposite to the most fundamental characteristic of all 'probability curves.' "[224] The observation that the few mosaic mutants that had been found were males gave Muller an insight into the mutational event itself. "In all these cases, where a recessive sex-linked mutant was involved, the mosaic fly was a male. This result does not mean that mutation does not occur in females. . . . The more reasonable explanation is that mutations occur similarly in males and females, but they occur in only one X chromosome at a time, and therefore recessive mutant genes cannot manifest themselves in the female sex, on account of the presence there of the dominant unmutated allelomorph in the homologous X chromosome. . . ."[224]

Extending the implication beyond this, "The conclusion that a mutation happens in only one X chromosome at a time implies that the agent which ordinarily produces the mutation must be extremely localized in its site of application. The previously known fact that usually only one locus mutates, and not the neighbouring loci of the same chromosome, might have been explained on the supposition that the influence at work was chemically specific, only affecting a gene of a given composition, but in that case it might be expected that the two similar genes in homologous chromosomes would both be affected. The fact that they are not shows that the immediate cause of the mutation is not a diffuse influence existing throughout the body, the cell, or even the nucleus. The mutation is due to an event of such minute proportions, so circumscribed, that it strikes only a single one of two nearby, similar loci in the same nucleus."[224]

Morgan, speaking before the Royal Society in 1922, did not carry this "prophetic vision" of the gene into his analysis. He remained close to the facts. "The chromosome theory, namely, that the hereditary units are carried by the chromosomes, means today much more than that the chromatin is the material basis of heredity. The theory carries with it the idea of the individuality and continuity of the chromosomes."[209] Crossing over as a mechanism was a simple physical model. "It seems to me that there is no escape from the conclusion that interchange of equivalent blocks of genes (i.e., pieces of chromosomes) takes place in a perfectly orderly manner for anyone who accepts the view that the 'chromosomes' are the bearers of the hereditary units."[209] As for the genes themselves, "The evidence from crossing over has led to the conclusion that the hereditary elements, the genes, are arranged in linear order in the chromosomes. If we think of these elements as material particles they must be supposed to have the power of self-division, and to remain unchanged through long periods. More than this we need not postulate. How they affect the cells in which they lie we do not know. Nor do we know whether they are functioning all the time, or only under specified conditions. However they do their work, we must regard the organism in large part as the outcome of the sum total of their activities."[209] The genes, whatever they were, were not mysterious units garbed with "philosophical" properties. "It is . . . important to emphasize that these ultimate units are not necessarily to be thought of simply as the representatives of each part of the organism, for every part of the organism must result from the activity of a large number of the elementary units. . . . I find that many people get the impression that we suppose that each part of the adult organism must have a single representative in the germ track, and sometimes the impression is produced, unfortunately, that we imagine the chromosomes as representing a sort of miniature of the animal. One of my friends has laboured for years under the impression that our idea of the germplasm is that of a mysterious insect in the nucleus, a sort of insectulum, and, under the circumstances, I cannot blame him for thinking that we are on the way to the madhouse."[209]

The size of the gene in *Drosophila* could be calculated if the total number of genes was accurately estimated. Using a method that "was first suggested by Muller," Morgan obtained 2000 genes for the gamete. The volume of the chromosomes at metaphase could be calculated from their dimensions (7.5 μ in length and 0.2 μ in diameter). The volume of the genes from this cylindrical model approximated spheres with a diameter of 0.06 μ. "The size of the gene on the basis of these tentative estimates seems to be larger, but not much larger than that of some protein molecules."[209] Accordingly, "the evidence has given us a glimpse at least of processes that are so orderly and so simple as to suggest that they are not far removed from physical changes; and the order of magnitude of the materials is so small as to suggest that its component parts may come within the range of molecular phenomena. If so, we may be well on the road to the promised land where biological results may be treated as physical and chemical events."[209] This was as far as Morgan would go.

The development of the gene concept and chromosome theory would now transfer gradually into the competent hands of Morgan's students. Embryology, the original inspiration for Morgan's detour into genetics, needed his attention. There was nothing known about the field of developmental genetics, but Morgan hoped to discover its principles in the marine forms he had studied in such detail. The gene concept had come to fruition through a circuitous route in Morgan's experiments. The nineteenth century philosophical units, rejected as mere speculation, played only minor roles. The chromosomes, through the discovery of "associative inheritance," acquired a factorial interpretation. The factorial hypothesis, in turn, forced continued characterization of the factors until they became genes. Twelve more years would elapse, however, before these "forced" conclusions would be recognized by the award of a Nobel Prize to their chief advocate.

The Individual Gene

Six months before Morgan presented this viewpoint in his Croonian Lecture to the Royal Society, Muller appeared in Toronto at a symposium on variation, the same meetings at which Bateson expressed his "conversion" to the main point of the chromosome theory. Muller's paper, "Variation Due to Change in the Individual Gene" is an electrifying document: it contains no data; it presents no new experiments; it presents ideas. "The present paper will be concerned rather with problems, and the possible means of attacking them, than with the details of cases and data."[225]

The significance of the gene had to be made known and Muller, with the fervor of Bateson some twenty years before, stated his case in unmistakably forceful terms. "This . . . fundamental contribution . . . to cell physiology . . . which has so far scarcely been assimilated by the general physiologists themselves, consists in the demonstration that, besides the ordinary proteins, carbohydrates, lipoids, and extractives, of their several types, there are present within the cell *thousands* of distinct substances—the 'genes'; these genes exist as ultramicroscopic particles; their influences nevertheless permeate the entire cell, and they play a fundamental role in determining the nature of all cell substances, cell structures, and cell activities. Through these cell effects, in turn, the genes affect the entire organism."[225] The exploration of the gene by crossing over had formed the boundaries of the factors, the study of the gene itself needed new approaches. "There is . . . another method of attack, in a sense more direct. . . . That is the method of investigating the individual gene, and the structure that permits it to change, through a study of the changes that occur in it, as observed by the test of breeding and development. . . . To understand the properties and possibilities of the individual gene, we must study the mutations as directly as possible, and bring the results to bear upon our problem."[225] Among these "attacks" would be the analysis of the structure of the gene. "It will even be possible to determine whether the entire gene changes at once, or whether the gene consists of several molecules or particles, one of which may change at a time. This point can be settled in organisms having determinate cleavage, by studies of the distribution of the mutant character in somatically mosaic mutants."[225] Another "attack" would be the estimation of mutation frequency. "In the past, a mutation was considered a windfall, and the expression 'mutation frequency' would have seemed a contradiction in terms. To attempt to study it would have seemed as absurd as to study the condition affecting the distribution of dollar bills on the sidewalk. You were simply fortunate if you found one."[225]

Most important, however, was the need to use fresh and exciting techniques, new organisms, new concepts. The viral elements, bacteriophage, found by Twort and d'Herelle, were examples of such new possibilities "which must not be neglected by geneticists." Their study might show whether they were some type of "autocatalytic" molecule or new organism. "On the other hand, if these d'Herelle bodies were really genes, fundamentally like our chromosome genes, they would give us an utterly new angle from which to attack the gene problem. They are filterable, to some extent isolable, can be handled in test-tubes, and their properties, as shown by their effects on the bacteria, can then be studied after treatment. It would be very rash to call these bodies genes, and yet at present we must confess that there is no distinction known between the genes and them. Hence we cannot categorically deny that perhaps we may be able to grind genes in a mortar and cook them in a beaker after all. Must we geneticists become bacteriologists, physiological chemists and physicists, simultaneously with being zoologists and botanists? Let us hope so."[225] At the end of Muller's talk, Henry Fairfield Osborn, believing he had been treated to a rare public display of scientific fantasy, congratulated Muller on his wonderful sense of humor.

Muller continued his conceptual analysis of mutation. In 1923 the term took on a specific definition. "The term mutation originally included a number of distinct phenomena, which, from a genetic point of view, have nothing in common with one another. They were classed together merely because they all included the sudden appearance of a new genetic type. Some have been found to be special cases of Mendelian recombination, some to be due to abnormalities in the distribution of entire chromosomes, and others to consist in changes in the individual genes or hereditary units. . . . The usage most serviceable for our modern purpose would be to limit the meaning of the term to the cases of the third type—that is, to real changes in the gene. . . . In accordance with these considerations, our new definition would be 'mutation is alteration of the gene.' "[226]

With this concept of mutation Muller raised "fourteen points" about the nature of the gene: (1) it was as stable as a radium atom with a probability of "decay" measured in a few thousand years; (2) genes showed differential mutability, some genes being highly mutable; (3) the external environment had negligible effects on changing mutation frequency and no effect at all on the direction or specificity of that change; (4) genes were not just losses, some could undergo reverse mutation to their original form or to other directions; (5) some multiple allelic series (the Truncate series) had qualitatively different effects; (6) some genes showed a preferential direction in the type of mutation expressed (the "skewed-curve" distribution of the white-eye series); (7) mutation frequency itself was gene-controlled and the mutation frequency could be altered by mutation at other loci; (8) only one specific gene at a time, as a rule, was affected in a cell; (9) mutation affected only one of two alleles in a nucleus—it was localized; (10) the gene could mutate at any stage of the life cycle of an organism; (11) the dominance expressed by normal genes was more complete than that expressed by newly arising dominant mutations; (12) the overwhelming number of gene mutations were deleterious to the organism; (13) most mutations involved slight rather than major modifications of the organism (they were Darwinian!); and (14) an entire spectrum of gene mutations existed ranging from lethal to visible, with the most frequent type being the recessive lethal.

Muller's "fourteen points" indicated the direction of his work over the next five years—research that would inevitably lead to a search for agents to induce mutations. The mutant gene would no longer be a "windfall" like a dollar bill in the street; it would become the geneticist's common currency.

CHAPTER 11

The Drosophila Group:
An Enigmatic Appraisal

The development of such a productive scientific concept as the "classical" gene theory deserves the detailed attention of historians. The gene concept, through the efforts of the "Drosophila group" gradually assumed a central role in biological thinking; it led in one direction to the regeneration of a mechanism for Darwinian evolution through the very biometric analysis of populations that originally resisted Mendelism; it led, in another direction, to the unity, through common genetic problems, of diverse biological fields, especially cytology, embryology, and cell physiology. Its most enthusiastic supporters, like Muller, visualized its eventual union with biochemistry, biophysics, and microbiology.

There is, however, no single interpretation of how this remarkably fertile concept was developed at Columbia University. Morgan visualized the process as a co-operative group effort between himself and his students. "It was not unusual for the six of us to carry on in this small room, the only space at our disposal. These were the days when bananas were used as fly food and in one corner of the room a bunch of bananas was generally on hand,—an adjunct to our researches which interested other members of the laboratory in a different way. As there were no incubators, a bookcase and a wallcase were rigged up with electric bulbs and a cheap thermostat, which behaved badly at times, with consequent loss of cultures. The use of milk bottles came into the program at an early date, but where they came from was not known, or at least not mentioned. At a later date we were bold enough to ask for a case of new bottles. . . . The picture of the conditions

under which we worked is not intended to suggest that we were handi-
capped. On the contrary, our proximity to each other led to cooperation in
everything that went on. The discovery of a new mutant was immediately
announced, and its location in the gene chain anxiously awaited."[211] The
"fly room" with its eight desks was a privileged sanctuary for a number of
Morgan's students; for some like Bridges and Sturtevant, it became a second
home for the next seventeen years; for others it was a temporary shrine
attained through awards of post-doctoral fellowships from abroad. Still
"others, who did not have desks in the 'fly room,' but who worked actively
with the group and were often in and out, are too numerous to mention
individually—but among them H. J. Muller must be especially important,
since his share in the early developments was especially important."[304] The
students, permanently or temporarily stationed in the room from 1910 to
1915, constituted "the group."

Sharing Morgan's view, Sturtevant presented an idyllic vignette of
scientific co-operation at its best. "This group worked as a unit. Each carried
on his own experiments but each knew exactly what the others were doing,
and each new result was freely discussed. There was little attention paid
to priority or to the source of new interpretations. What mattered was to
get ahead with the work. There was much to be done; there were many
new ideas to be tested, and many new experimental techniques to be developed. There can have been few times and places in scientific laboratories
with such an atmosphere of excitement and with such a record of sustained
enthusiasm. This was due in large part to Morgan's own attitude, compounded of enthusiasm combined with a strong critical sense, generosity,
open-mindedness, and a remarkable sense of humor. No small part of the
success of the undertaking was due also to Wilson's unfailing support and
appreciation of the work—a matter of importance partly because he was
head of the department. . . . Because of the close cooperation in the work
it is very difficult to trace the individual contributions to the developments
in this period."[304]

A less sanguine view of the "group" was expressed by Muller in 1934.
"It was in fact in the hope of gaining evidence for his various substitutes
for Darwinism that Morgan began in 1909 his experiments with *Drosophila*.
. . . The results proved, however, to be glaringly at variance with his views,
and at the same time he soon found himself pressed in his interpretations,
by a small group of younger co-workers occupying the official positions of
'students,' whose ideas, despite their officially subordinate position, Morgan
realized that he should take seriously. These 'students' had been influenced
greatly by their studies under Wilson, and even more by Lock's remarkably
prophetic book ('Recent Progress in the Study of Variation, Heredity, and
Evolution,' 1906, 1909) which has precisely the modern standpoint upon
all essential questions of heredity (role of the chromosomes and their interchange of linked genes, universality of Mendelism, multiple factors, and
even something of 'balanced lethals'). Slowly, and against his will Morgan

was forced to give way to this double pressure of facts and arguments. So far as concerns the role played here by the winning of the facts themselves, it may be remarked that the earlier of these, in 1909–11, were mostly contributed by himself, while the great bulk of the facts of real significance subsequent to 1911, and practically all after 1913, were found by the younger workers quite independently of any guidance from him, in experiments which they had planned on the basis of their own more advanced viewpoints. Their results and interpretations were, however, later accepted by Morgan and presented chiefly by him to the scientific and lay public, so that these developments have sometimes been referred to, especially in circles farthest removed from contact with the original work, as 'Morganism.' In this way did the central trend to modern materialist genetics have its origin."[234]

Muller's main point, that Morgan became a geneticist "in spite of himself" is amply illustrated by the critical attitude he expressed prior to 1910 against Sutton's chromosomal interpretation of Mendelism, against the purity of the Mendelian factors, against the logical, rather than experimental, basis for factors and their localization in the chromosomes, and against the more naïve theories of Darwinism. This attitude of Morgan's, expressed in detail in 1909 in his address "For Darwin" is a mixture of respect for "the spirit of Darwinism" and an attempt to replace the "fluctuating variations" of Darwin with the "definite variations" or mutations of all magnitudes, that de Vries had proposed for speciation. Not selection, but the filling in of new ecological niches through the breeding of new "definite variations" was favored in Morgan's view. Whether Muller is correct that Morgan's interests in Drosophila were initiated by "the hope of gaining evidence for his various substitutes for Darwinism" is difficult to evaluate. Morgan did not dispute the occurrence of evolution; he disputed the Lamarckian and biometric interpretations of the mechanism of evolution. Unfortunately, both these interpretations favored that view of minute imperceptible differences that later on, some twenty years later, would find their support in the factorial hypothesis that Morgan helped to establish. At the time, however, Morgan voiced skepticism. It is not clear what Morgan hoped to do in his attempts to induce mutations in Drosophila. The high frequency of mutations in Oenothera would have rapidly pointed out the lack of a parallel mechanism in Drosophila. Morgan, however, persisted. According to Sturtevant, "Morgan began work in Drosophila in an attempt to induce mutations; but before he took up that material he had already begun his strictly genetic work using mice (beginning in 1908) and rats (1909)."[304] Morgan (1914) also alludes to a quest for mutations in these first experiments with Drosophila. "One of the first mutants that I observed in ampelophila appeared in the offspring of flies that had been treated with radium and although there was no proof that the radium had had a specific effect I felt obliged to state the actual case, refraining carefully from any statement of causal connection. Nevertheless, I have been quoted as having produced the first mutant by the use of radium. I may add that repetition of the experiment on a large scale both with the emanations of an x-ray machine and from radium salts has failed to produce any mutations, although the flies were made sterile for a time."[205] Morgan was, however, disappointed with the progress of his *Drosophila* work just prior to the discovery of the white-eyed mutant. Whether this was frustration at not finding proof for a "substitute" for Darwinism or a failure to induce variations for use in genetic studies is not known. Ross G. Harrison, who had been a graduate student along with Morgan in Brooks's laboratory, had visited Morgan during this period early in 1910. "I recall a visit to Schermerhorn Hall, about 1910 or at least before the Drosophila visitation. Morgan waved his hand at rows of bottles on shelves and said: 'There's two years' work wasted. I've been breeding these flies for all that time and have got nothing out of it."[162]

The influence of Lock's text is unquestioned; but Lock did not derive his ideas on the universality of Mendelism, multiple factors, and 'balanced lethals' from his own reflections and insight. He got these from Bateson's 1902 report to the Evolution Committee. Of this conclusion there can be no doubt from a reading of Bateson's penetrating analysis of what the new discovery meant for the future of biological science. Lock himself virtually eulogizes Bateson's influence on his thinking. "Adequately to acknowledge Mr. Bateson's influence upon these pages is a more difficult matter, and not the less so because I have deliberately refrained as far as possible from consulting him whilst the book was in course of preparation, in order that it might retain if possible some traces of individuality. It is therefore clear that he is in no way responsible for its deficiencies. But apart from the fact that I am conscious of having quoted his ideas at more points than could possibly be acknowledged *seriatum*, I owe to Mr. Bateson both my first introduction to the science of genetics, and a continual fund of encouragement in the prosecution of studies connected with it."[188]

What Lock achieved, however, was the very union of cytology and Mendelism which was later borne out at Schermerhorn Hall. The two fields which Lock integrated were brought together, not on the basis of any one critical experiment (such as Morgan was to do four years later), but on the basis of de Vries's modifications, in 1903, of his theory of intracellular pangenesis, as well as Weismann's early ideas on the continuity of the germ plasm and gamete formation. It was Bateson's deficiency that the panorama of implications of Mendelism was based exclusively on breeding analysis. In this regard an interesting parallel can be made of the effect of a new discovery on its first recipients. Morgan, in his address "For Darwin" points out the paralysis in fruitful research that followed Darwin's publication of *The Origin of Species*. "So extensive were the facts of variation accumulated by Darwin, so penetrating was his analysis of these facts, so keen was his insight, and so wise his judgement as to their meaning, that for thirty years afterwards little of importance in this direction was added. In their amazement at Darwin's accomplishment zoologists forgot that he had opened the door leading to an unexplored territory."[197] So, too, in the amazement at Mendel's discovery, Bateson displaced any other approach save Mendelism itself. This had its roots, of course, in Bateson's scientific development. Bateson, raised in the descriptive embryological school of Lord Balfour at Cambridge, then succeeding in an embryological analysis of *Balanoglossus* (which he later considered just a "trifle"), returned to that same descriptive laboratory environment. For most of his life this unproductive "Darwinism-through-descriptive-embryology" attitude was distastefully associated with microscopy. Similarly, there was a prolific number of speculative theories on the role of the chromosomes, with accompanying hypothetical units; this habit was characteristic of the German school of cytology and embryology. These theories, void of experimental support, were also associated in Bateson's outlook with the inadequacy of microscopy as the tool for studies in heredity. Later in life, when he visited Morgan's laboratory in 1921, he acknowledged this weakness. "Cytology here is such a commonplace that every one is familiar with it. I wish it were so with us."[24]

In the critical years, about 1906, Bateson's attitude was therefore one of skepticism. "The recognition of a definite differentiation among the chromosomes (see especially Sutton 1902) is probably an important advance, though until we can positively recognize characters in the zygote as associated with some visible cytological elements we must be cautious in forming positive conclusions as to the relation of cytological appearances to the phenomena of heredity."[17]

On Edmund B. Wilson's influence there is agreement between Sturtevant and Muller. Morgan and Wilson enjoyed their association at Columbia, "The two men did not always agree on scientific questions, but the disagreements were openly discussed and each respected the other's opinions."[304] Both were students of Brooks, although the cool attitude of Morgan towards Brooks was the reverse of the warm regard which Wilson held for him. "It was through informal talks and discussions in the laboratory at his house, and later at the summer laboratories by the sea that I absorbed new ideas, new problems, points of view, etc. . . . Through him I first discovered what I really wanted to do. . . . From him I learned how closely biological problems are bound up with philosophical considerations. He taught me to read Aristotle, Bacon, Hume, Berkeley, Huxley; to think about the phenomena of life instead of merely trying to record and classify them."[212]

Wilson enjoyed the cellular analysis of development as much as Bateson disliked it. He worked out cell lineages in a variety of invertebrates in the hopes of resolving the controversy between Roux and Driesch on the capacity of cleavage cells for generating complete embryos. In the course of these studies he became convinced that the resolution would come through cytology itself and not through cell lineage studies alone. The cell theory for Wilson was bound up with the problem of heredity and the German school appealed to him. Towards the end of the nineteenth century he summarized these views in *The Cell in Heredity and Development*. The book had a profound influence on the development of the chromosome theory of heredity. The book was only part of Wilson's invigorating effect on the "Drosophila group" that was to arise at Columbia. Muller's evaluation gives a vivid impression of that climate of enthusiasm which Wilson conveyed to the undergraduates at Columbia.

"The excitement of the advances in chromosome theory made by Wilson in 1905–10 communicated itself through the department of zoology at Columbia. This helps to explain why it was that most of the first batch of youngsters who became *Drosophila* workers with Morgan had been undergraduates at Columbia College in the latter part of the period and that others came through by the same route soon afterwards. Lured on by their first course in biology, where they were molded by Sedgwick and Wilson's text and by the teaching of Calkins and McGregor, both former students of Wilson's, some of them had the privilege of taking in their sophomore year Wilson's thrilling one-semester course on heredity and the chromosomes, variation, and evolution. In this the text chosen by Wilson was Lock's extraordinary book of 1906—too far 'ahead of its time' to be sufficiently remembered now—which, with less caution and fewer qualifications than employed by Wilson himself, advocated the sufficiency of Mendelism, multiple factors, the chromosome theory (including exchange of linearly arranged genes during parasynapsis, after de Vries) and the natural selection of mutations as the basis of all heredity and evolution. Wilson's superb course on cytology, with its unequalled laboratory training and demonstrations, was usually taken by them in their third or fourth year after entering as Freshmen. After this stimulating and thoroughly systematic preparation their embarkation upon the adventure of the fascinating new work on chromosome heredity in *Drosophila* that had just been opened up by Morgan (1910 and 1911) was the logical continuation, now grown more specific in its direction, of the quest to which they had already become dedicated, calling for the ways of thinking, the knowledge and to some extent even the technique acquired during their previous years of training. And the striking similarity in the attitude of all of them towards the new problems was in no small measure a reflection of the degree to which this common training had been driven home. Thus it is likely that only these *Drosophila* workers, of the earliest years, fully realize to what an extent modern genetics traces its descent through Wilson."[242]

The contrast in evaluation between two of Morgan's own students is a striking one. To some extent it is an artificial distinction since Sturtevant's views were expressed in an obituary for Morgan and Muller's were contained in an article aimed at an unrelated issue. Obituaries tend to minimize the deficiencies and inadequacies of a great man and bring forth the heritage of his work which serves as his epitaph. Muller's critique, written while

Morgan was alive (the year Morgan received the Nobel Prize in Medicine and Physiology) appeared in a philosophical context with the title "Lenin's Doctrines in Relation to Genetics." Written at a time when formal genetics was being attacked in the U.S.S.R., the article reads "between the lines" as a counter-attack on the rival genetic school in the U.S.S.R. which supported a neo-Lamarckian non-Mendelian theory of heredity and evolution. The controversy, which had its roots in the 1920's eventually erupted into a clear cut "Lysenkoist" camp and a formal genetics camp, disparagingly referred to as "Morganist." "Morganism," to this rival school, represented "reactionary" and "idealistic" tendencies which ran counter to the "materialist" attitudes expressed by Marxian and Leninist (later Stalinist) outlooks on science. The controversy eventually forced Muller to leave the U.S.S.R. a few years after the writing of this article. The article itself dissociates "Morganism" from the gene concept developed at the Columbia laboratory, and in doing this, Muller pointed out the indecisiveness, reversal of attitude, and reluctance that Morgan expressed about the chromosome theory before 1910. Muller's interpretations, cast in a "dialectic" phraseology, were obviously intended to placate governmental (and popular) fears that formal genetics was "bourgeois." The battles over the unit-character fallacy and the presence and absence hypothesis were used, not as examples of less productive theories, but as illustrations of "idealistic" or "muddled" philosophy. The participants in these controversies, especially Morgan, were offered as "whipping boys" to appease the potential supporters of the politically aggressive foes of formal genetics in the hopes that the intrinsic merits of the gene concept and chromosome theory would be permitted to dominate genetic research in the U.S.S.R. The lumping of all the genetic controversies, (especially the various theories of the gene which had been espoused by all the diverse personalities since 1900) provided an easy target for attack by the anti-Morganist school. Muller's defense of the gene concept, unlike Bateson's defense of Mendelism, failed because (as Muller learned with sorrow) the debate was more than a contest between two rival genetic theories.

Whether this interpretation of this curious document is correct, or whether it is a charitable rationalization may never be known. The personal feelings and evaluations expressed in one generation may strike an entirely different connotation in a later generation. Bateson, for example, will long be remembered for the defense of Mendel, not for the personal feud with Weldon and Pearson. For scientists, his aphorisms are frequently used as guidelines of the research attitude they champion: "If I may throw out a word of counsel to beginners, it is; Treasure your exceptions! When there are none, the work gets so dull that no one cares to carry it further. Keep them always uncovered and in sight. Exceptions are like the rough brickwork of a growing building which tells that there is more to come and shows where the next construction is to be."[18] This view is in marked contrast to a less frequently quoted opinion. "Democracy is the combination of the mediocre and inferior to restrain the more able."[21] The statement was made in 1919 in a speech which attacked nationalism and war (he lost a son in World War I) and which opposed the formation of a League of Nations on the grounds that it was based on nationalistic and not international principles.

Morgan's role in the development of the gene concept remains enigmatic. He fought it, contributed to it, grew to believe it, fought hard for it, and reaped the rewards for conveying it to the world. Morgan's willingness to accept experiments which contradicted his own theories and predilections was important for the development of the gene concept. His uncompromising rejection of Bateson's coupling and repulsion theories on the basis of his analysis of white, miniature, and yellow was essential for the progress of genetics but its brusqueness led to an unfair rejection of Bateson's earlier contributions to genetics. Morgan became the spokesman for the new genetics; his students found, through him, the public attention which their own work would not have created on such short notice. Morgan also provided an atmosphere in Schermerhorn Hall and later at the California Institute of Technology in which the more productive of his students could work. Even if he did not play a major role in the discoveries of mapping, interference, non-disjunction, and residual inheritance, he acknowledged their significance and incorporated them into a broader view of chromosome theory.

If Muller, Bridges, and Sturtevant had not been there, others might have been. The progress might have been slower, the directions more devious, but the localization of factors would have been forced upon them (as Morgan's 1912 analysis of the lethal factor near white suggests), and the subsequent maps, in turn, as they became crowded, would have forced the factors themselves into acquiring the characteristics of genes.

CHAPTER 12

The Genomere Hypothesis

While the gene concept was becoming transformed into a material reality for Morgan—the reality of the particulate unit—it was viewed in a novel way by Goldschmidt. Citing his own research on the gypsy moth *Lymantria dispar*, Goldschmidt, in 1917, claimed that the "relative sexuality," the degree of maleness and femaleness in different races of this genus, varied in a quantitative series similar to that of the multiple allelic mutants of the white-eye locus. This was even more striking for certain patterns, degrees of melanization, in the nun moth, *Lymantria monocha*. Although no crossover analysis was used in arriving at his conclusions, Goldschmidt accepted a multiple allelic interpretation for these cases. "A careful consideration of these points shows clearly what these multiple allelomorphs for pigmentation really are: they are different quantities of the substance which we call a gene which act according to the mass-law of chemical reactions, i.e., produce a reaction or accelerate it to a velocity in proportion to their quantity."[141] This interpretation of the gene provided a physiological basis for Castle's belief that the gene was a fluctuating unit subject to contamination. "If our conclusions regarding the nature of multiple allelomorphs are accepted, it must lead to a different intellectual attitude toward the problem of variability of genes, which is so important for evolution. . . . In the long controversies of recent years regarding the interpretation of Castle's work the logical side of the case seems to have been in the foreground. . . . We believe that this intellectual attitude toward the problem is the result of Johannsen's doctrine of agnosticism in regard to the nature of the gene, which resulted in a kind of mystic reverence, abhorring the idea of earthly attributes for a gene. . . . If however, it can be proven that genes are substances with the attribute of definite mass, it would be illogical to deny their variability. Nobody will claim that a gene is a substance that passes unaltered from generation to generation."[141] This material attribute was distinctly different from the material unit formulated by Morgan. "If now, the substantial basis of heredity in the sex-cells is established by the assembling of all factor-substances in their characteristic quality *and their correct quantity*, the situation is the same for the gene as for any other organic process; the varying conditions of the surroundings of the gene cause a certain amount of fluctuation in its quantity."[141]

Goldschmidt attributed the physiological mechanism for unit-character modification to reaction velocities of the genic materials. "The somatic character . . . can only change toward a plus or minus side. This change, caused directly by a difference in the velocity of reaction, however, can be produced either by the action of the medium, and then it is a modification, or by fluctuation in the quantity of the gene, causing increase or decrease in the velocity. . . . The *deus ex machina* modifying factor, which moreover, does not fit the decisive genetic facts in the most discussed case of Castle's rats nor our cases, thus becomes superfluous."[141] The critical analysis of Beaded in *Drosophila*, and a similar analysis of eosin and truncate, showed, however, that factor modification in *Drosophila* was demonstrably caused by the "*deus ex machina*" which Goldschmidt and Castle rejected. So, too, was the residual heredity in rats that Castle had to assume for his later crosses in the hooded series. While Goldschmidt's model was discredited by the success of the factorial hypothesis, it served as the progenitor in modified form of similar "quantitative" models of the gene.

The need for such a model was not long in coming. Correns, studying striped and variegated leaf coloration in *Capsella*, discovered that this pattern could be transmitted as a Mendelian character. Correns considered this instability of the character a property of "sick genes." In 1919, Correns proposed a model for these anomalous genes. "In order to have at least a model one might assume that the gene consists of a large molecule to which the same side chain of atoms is attached, say ten times. This number might be mutable, might undergo changes in the plus or minus direction under unknown conditions 'external' for the gene. To each number of these side chains would correspond a definite ratio of white and green in the mosaic plant. . . . The difference in this interpretation . . . would be that the state of the gene, the number of side chains attached to the gene molecule, is not constant, but that new chains could be added or old ones detached, and this during ontogenesis of the individual."[88]

Maize geneticists, particularly R. A. Emerson and his students at Cornell, were interested in similar problems of variegation. The character they studied was expressed by members of a multiple allelic series affecting coloration of pericarp in the kernels and in the cob of the ears. "There are in this somewhat remarkable series of multiple allelomorphs, many degrees of factorial constancy. On the one extreme are self-colored and colorless races, both as constant probably as most Mendelian characters. Next to

these are the self-colored types that exhibit from one to four or five variegated or partially variegated seeds on perhaps a majority of the heterozygous
ears. Then comes the light variegated type in which a factorial change to
self color comparatively rarely occurs at a sufficiently early stage in ontogeny to affect the germ cells but somewhat more frequently at later stages.
At the other extreme is a little known type of very dark variegation in which
the factorial change occurs so frequently that all ears, so far as observed,
have numerous self-colored or near self seeds. . . . While the factorial
changes described are doubtless non-Mendelian, they are in no sense anti-
Mendelian. Genetic modifications are not the concern of Mendelism but
of mutation. The essential feature of Mendelism is the segregation of unit
factors without their contamination."[124] This view, while adopting mutation as a descriptive or suggestive term did not provide a model for the
origin of mutations leading to the variegation and spectrum of types.

Three years later, in 1920, a model was suggested by E. G. Anderson
in a letter written to Emerson. "In order to explain frequent mutations in
unstable genes he suggested that genes are not the ultimate units of heredity
but are compound structures composed of still smaller units. In the case of
variegated pericarp color he assumed that unstable genes contained two
kinds of sub-gene particles, a white-determining kind and a red-determining
kind, which by mere segregation formed either white- or red-determining
genes."[104] The theory was publicly announced in 1921 and extensively developed in 1924 by W. H. Eyster. Eyster began his study with a single ear of
orange seeded maize and he obtained, from the progeny ears, the entire
spectrum of coloration that Emerson had reported for the whole series of
alleles. He also found that the variegated forms could be selected in either
direction and that white or red seeded ears could be derived from them as
stable forms. To account for these results, Eyster extended Anderson's suggestion that a compound gene was involved. "The changes or variations in
the gene for orange pericarp form an orthogenetic series of colors ranging
from whitish to deep cherry-red. Such a quantitative variation can be most
easily explained by assuming that the gene for orange pericarp is a compound structure, made up of pigment producing *gene elements* and non-
pigment producing *gene elements*. According to this view the intensity of
color depends upon the relative numbers of the contrasting gene elements
incorporated in the structure of the gene. An excess of non-pigment-
producing gene elements would produce a light orange color, while an
excess of the pigment-producing gene elements would produce a deep
orange or red color."[127] Occasional seeds of different hue in an otherwise
uniform ear could be attributed to the distribution of the gene elements.
"Because of the random assortment of gene elements the existing relation
between the contrasting gene elements would be maintained in a relatively
large number of plants. In a somewhat smaller number of plants there
would be expected a slight excess of the pigment-producing or non-pigment-
producing elements in the tissue in which the color is expressed, due to

slight inequalities in the separation of the elements at the mitoses, or to a slight difference in the rate of reproduction of the individual gene elements."[127]

This readily accounted for the hue of the various seeds and ears but it did not satisfy the variegation phenomena. "The fundamental difference between orange pericarp and variegated pericarp is that in the former the pigment is diffused throughout the tissue, while in the latter the pigment is limited to segments and patches of various sizes. In the gene for orange pericarp the complete segregation of unlike gene elements usually occurs not at all, or only very late in development, so that only small segments of red and colorless tissues are produced. In the gene for variegation the segregation of unlike gene elements begins early enough so that they are completely segregated by the end of pericarp tissue development. . . . It seems evident that a color variegation is produced when a gene includes both pigment-producing and non-pigment-producing *gene elements* which become segregated in the somatic tissue, in the course of development, by the mechanism of mitosis."[127] These gene elements, or "genomeres" as Eyster later called them, presented a new concept of gene structure, giving its proponents a new experimental approach. "This hypothesis, carrying the concept of the mechanistic structure of the gene to the extreme, was a definite forward step in gene research, since it opened a way for experimental approach to the problem. In the case of variegated pericarp, the red spots are of different sizes varying from very minute dots to large sectors covering a number of kernels. It seemed clear that . . . the differences in size must be due to differences in time of origin, namely that the large spots originated early in development of the ear and small ones at a later stage. Thus by classifying spots according to their size and determining the frequency of each class it was possible to find out the approximate rate of change in the successive cell generations. Also, by assuming that sub-gene particles assort freely it was possible to compute the theoretical expectancies for the rate of change at each division. Since that rate depends on the number of segregating particles within the large unit, an estimate of the number of sub-genes within an unstable gene could be made."[104]

Genomeres in Drosophila

The group of maize workers absorbed in the analysis of the variegation problem included Emerson, Anderson, Eyster, and M. Demerec. Demerec, dissatisfied with the long generation time required for maize experiments, turned his attention to a different organism, *Drosophila virilis*, in the hopes of exploring the problem of gene structure. His hopes were soon rewarded. In rapid succession, over a three-year interval, from 1924 to 1926, three genes were found which exhibited a type of variegation or instability that resembled the somatic and germinal mutability of unstable genes in maize.

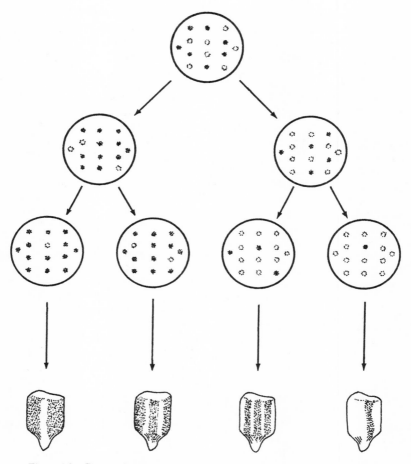

Figure 16. Demerec's Genomere Model. The variegation gene is assumed to contain a mixture of dark and light genomeres. Mitotic divisions lead to unequal distributions of light and dark genomeres. The phenotypes of the kernels derived from gametes bearing such unequal distributions reflect the proportions of genomeres present in the gene.

One of these, reddish-α, was an allele of the yellow locus in *D. virilis*. It gave frequent non-yellow progeny which were stable, somatically and germinally. Two other mutants, magenta-α and miniature-α were unstable, germinally and somatically. Some lines would maintain this variegation in all their progeny, other lines would give progeny that were either variegated like their parents or stable normal mutants. From these various cases, Demerec presented a detailed support of the genomere hypothesis. "A working hypothesis which assumes that the gene is not the smallest hereditary unit but a complex structure composed of smaller units would very well explain the conditions described. A complex-gene hypothesis was formed by Correns (1919) and E. G. Anderson (1921) to explain the mutable conditions on genes causing variegations in *Capsella* and maize respectively. The evidence presented here on the origin of mutations tend to generalize that hypothesis."[97]

This view was challenged by Muller in 1926. In an article which may be considered the apotheosis of the gene concept ("The Gene as the Basis of Life") Muller considered, among other topics, the implications of the genomere hypothesis. If a gene were a single unit and the mutational event required one or more replications for its fixation, then the pattern of variegation or mosaicism would be distinct from a gene composed of a number of similar sub-units. "The sorting out of descendants of a mutant gene-particle into just one block of cells (consisting of cells all more closely related to each other than to any others) would mean that the individual chromatin *strands*, at any rate, had a non-compound gene structure. A patchwork, on the other hand, would mean a compound gene, provided the cells remained together in order of their relationship."[227] The two models could be studied among the somatic mutations in *Drosophila* which had been found by the Columbia group. "A survey of the rather meager number of somatic mosaics in *Drosophila* which had originated by gene mutation then indicated that in all these cases the mutated region seemed to form one complete block, contrasted with the complete block of non-mutated tissue. The clearly bilaterally divided yellow and gray male reported by me in 1920 is a good example of this. This, then, makes it seem likely that, in *Drosophila*, the genes in each chromosome strand are units, in that they do not contain more than one of any given kind of interchangeable, separately mutable, self-propagating molecules (or particles of any order)."[227] Demerec's work, however, was still unexplained, and it was not known if the two classes of mosaic flies were attributable to one or several mechanisms. "It might be supposed that there is a partially orderly distribution of 'gene elements,' which only sometimes becomes irregular. Meanwhile, we must be prepared to watch closely for as many mutational mosaics as we can find, as well as to follow the possible segregations of the gene particles from generation to generation of individuals—though previously it had not seemed to me likely that such segregation would still remain incomplete after an entire zygotic generation. It is important to know whether these cases afford a real glance into the interior of typical genes or are—if I may so put it—'teratological.' "[227]

A more definite decision could be made in 1927 for *Drosophila melanogaster*. In that year Muller had discovered the "Artificial Transmutation of the Gene" through x-radiation. Among the numerous findings announced in this first paper on artificially induced mutation was one which ruled out the genomere model for most, if not all, of the somatic mutations arising in the experiments. "It is found that the mutation does not usually involve a permanent alteration of all of the gene substance present at a chromosome locus at the time of treatment, but either affects in this way only a portion of that substance, or else occurs subsequently, as an after-effect, in only one or two or more descendant genes derived from the treated gene. . . . This would imply a somewhat compound structure of the gene (or chromosome as a whole) in the sperm cell. On the other hand, the mutated tissue is distributed in a manner that seems inconsistent with a general applicability of the theory of 'gene elements' first suggested by Anderson in connection with variegated pericarp in maize, then taken up by Eyster, and recently reenforced by Demerec in *Drosophila virilis*. . . . A precociously doubled (or further multiplied) condition of the chromosome (in 'preparation' for later mitosis) is all that is necessary to account for the above-mentioned *fractional effect* of x-rays on a given locus; but the theory of a divided condition of each gene, into a number of (originally identical) 'elements' that can become separated somewhat indeterminately at mitosis, would lead to expectations different from the results that would have been obtained in the present work. It should, on that theory, often have been found here, as in the variegated corn and the eversporting races of *D. virilis*, that mutated tissue gives rise to normal by frequent 'reverse mutation'; moreover, treated tissues not at first showing a mutation might frequently give rise to one, through a 'sorting out' of diverse elements, several generations after treatment. Neither of these effects was found."[228]

This interpretation was confirmed for the lethals themselves in subsequent radiation experiments that year. Muller analyzed the F_3 and F_4 progeny sampled from the originally treated parental sperm in the hopes of

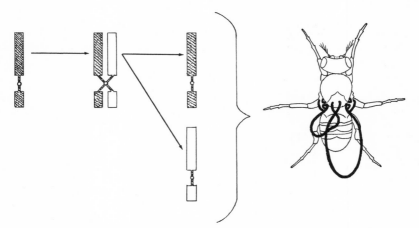

Figure 17. Muller's Fractional Mosaicism. In the course of analyzing spontaneous and x-ray induced mutation, Muller suggested that the chromosome contained two elements or that it was "precociously split." The mutational event frequently resulted in a fractional phenotype; presumably the mutant and non-mutant chromatids from the altered chromosome gave rise to the portions of mutant and non-mutant tissue in the fly.

finding a germinal sorting out of new lethals. A small number, probably of fractional origin, were found in the F_3 but none in the F_4. Neither somatic mosaicism nor germinal mosaicism in *D. melanogaster* was comparable to the variegation expressed by the mutable genes of *D. virilis*. In 1928 a radiation-induced "mottled" allele of the white region did show an "ever-sporting" character. "The continual mutations of mottled backward and forward, even in the somatic tissue, are indicated by the splotches on the eye, and different strains of it, having more and fewer spottings, and varying intensities and extensions of color in the spottings or in the background, have been established by selection. The theory of a compound gene might at first sight seem to find a very favorable case here, but it must be recalled that other allelomorphs of this gene are known (eosin, tinged, etc.) which, although they too show a color intermediate between red and white, are not unstable at all. Thus the instability of mottled would seem to depend on some other feature of the gene structure than the apparent mixture of color elements."[229]

Demerec, however, was not convinced that the genomere had been eliminated. "The results of the experiments obtained so far, can be explained by the assumption that the mutable gene is complex in structure."[99] However, "unquestionable acceptance of any hypothesis might do more harm than good to the progress of the work on this problem."[99] Similarly, Goldschmidt in 1928 was not discouraged by the criticisms of the genomere concept and "factor variability"; his own "quantitative" model even appeared strengthened by these experiments. "Normal, stable genes have the elementary property of being 'absorbed' in the chromosome, always in their characteristic quantity. . . . Unstable genes, properly called by Correns sick genes, lack something in their physico-chemical properties necessary for constancy in their number of molecules. . . . No different types of genomeres with quantitative assortment are needed, but simply the definite quantities of the one gene substance reckoned in numbers of molecules. . . . Thus the facts about unstable genes seem to add new material to the already broad basis of our general conclusions."[143]

Emerson accepted mutability *per se* as equally valid an interpretation of unstable genes as the genomeres for corn variegation. "The most plausible interpretation of the results . . . is that a gene or perhaps more than one, at a locus other than that of the variegation gene and its allelomorphs but in the same chromosome, influences the mutability of the variegation gene. . . . The difference between low grade and high grade variegation is merely that, in the one case, mutations from white to red occur less frequently than in the other. Or, if the hypothesis of gene elements is preferred, it is a matter of the relative rapidity of the sorting out of gene elements, dependent in turn probably on the particular combination of gene elements originally present."[125]

By 1931 Demerec pursued the study of mutable genes in two major directions. He attempted to obtain increases in the number of mutable genes in *D. virilis* and in their mutability through the use of radiation. Neither event was obtained. "This finding suggested that the reactions responsible for changes in unstable genes are not increased by x-rays in the same proportion as reactions responsible for changes in stable genes."[104] Turning to the variegation of flower color in *Delphinium*, where the sectors were clear-cut and quantitatively measurable, Demerec hoped to calculate the number of genomeres involved from the pattern of variegation. With the assistance of Sewall Wright, Demerec constructed a predictive model of genomere segregation to account for the variegation patterns obtained in each generation. "The results obtained with *Delphinium* indicated strongly that the mechanism responsible for variegation was not as simple as was assumed by the subgene or genomere hypothesis. . . . Data on the mutability rate of unstable rose flower color showed that gene to be mutating with approximately the same rate throughout the twelve cell generations covered by these data. These observations did not agree with the theoretical expectancies based on Wright's calculations. According to Wright, in order to obtain the constant rate of change of 7^{-5} per cell generation which was observed in unstable rose, about 4000 genomeres would be required. With such a high number of components, thousands of cell generations would be needed before the constant rate of change would be approached."[104] Abandoning the genomere hypothesis, Demerec favored Emerson's interpretation: "It appears probable that changes in unstable genes are rather of a chemical nature."[104]

For the geneticist in 1931 there were now three classes of variegation: somatic and germinal "fractional" mutation as in *D. melanogaster* attributable to the structure of the "chromatin"; mutable genes characteristic of plants and *D. virilis*; and "eversporting" mottled mutations in *D. melanogaster*. Their ultimate interpretations would require new techniques and concepts. At the beginning of the 1920's the discovery of unstable genes posed a threat to the "classical" gene concept, but at the close of the decade the gene concept remained unitary.

CHAPTER 13

Position Effect: Bar Eyes

In 1913 a single male with a striking abnormality was observed by S. C. Tice in a bottle of flies. The eyes of the male were reduced to thin bars. Analysis of the mutant showed it to be sex-linked and dominant to its normal, round-eyed, allele. When homozygous, the Bar eyes in females looked identical to those in the Bar males, but heterozygous Bar females, B/+, showed a broadening of the width of the Bar with an increase in the number of facets (ommatidia) in the compound eyes. It was thought that the normal allele was not completely recessive to Bar and contributed to the increased number of facets. This interpretation of Bar was held until 1917 when Bridges discovered the first case of a "deficiency" of genes. When this deficiency (which lost the Bar gene and the gene for forked) was rendered homozygous it was found to be lethal. When crossed to a Bar-eyed male, however, the compound of the deficiency and Bar showed the same moderate Bar eye that was present in the heterozygote of Bar with its normal allele. Bridges was startled by this observation. "It was concluded that the broadness of the eye of the normal heterozygous Bar female is not due to the action of the wild-type allelomorph, as formerly supposed, but to the fact that one narrowing gene (B) is opposed by two sets of broadening genes, one situated in each of the X-chromosomes in some other region than that involved in (f-B) deficiency."[215] It seemed then, that Bar had arisen as a "new presence" with no allelic counterpart.

A few years earlier, in 1914, Charles Zeleny wrote to Morgan for a sample of this stock. He used it for selection studies to see if he could broaden and narrow the size of Bar eyes. In this he was successful and it seemed that the selection was building different modifier genes for the plus or minus direction of selection. To assist him in this study, Zeleny urged

H. G. May to work on this project. To his surprise, May found that among some 9000 flies that he observed, an unexpected number of them had reverted to round eyes. "During some experiments in selection for higher and lower facet numbers in the bar-eyed race of Drosophila ampelophila, I obtained six full-eyed males and five heterozygous females from the stock bottles and the selected lines."[191]

Zeleny pursued May's observations and amassed a sizable number of observations on reverse mutation at the Bar locus. In 1921 he reported that no forward mutations to Bar were obtained from round-eyed normal flies among 46,000 progeny. But the reverse mutation frequency was phenomenally high: 53 round-eyed flies appearing among 85,000 Bar progeny. Additionally, Zeleny discovered a more extreme variant of Bar with such a narrow slit that it was often broken up, presenting one or two traces of ommatidia on each side of the head. He called these mutants "ultrabar" and found their frequency among the Bar progeny was much lower than the occurrence of round-eyed types. Only 3 ultrabar flies were found among the 85,000 progeny with Bar eyes. Ultrabar, like Bar, was unstable and both round-eyed (5/8700) and Bar eyed (3/8700) flies were obtained from it.[322]

When May had found his eleven exceptions he concluded that the mutations arose only among the females because the 1 : 1 ratio was higher than what would be expected for the reversions coming from both sexes (in which the ratio would be 2 females : 1 male for the reverted exceptions). Zeleny was not convinced they were exclusively from females but his ratios were closer to 1 : 1 than 2 : 1. Among all his series, there were 33 changes affecting females and 29 changes affecting males. For the single series of 85,000 progeny from Bar, 31 were females and 20 were males among the testable exceptions. Another peculiar feature, in contrast to Muller's analysis of the origin of mutations at the white locus, was the paucity of mosaics that appeared in the reversions, either somatically or gonadally. Yet as a dominant mutation, many more mosaics should have appeared, at least somatically, than for recessive white. "Of the forty-four separate mutations which are suitable for the present purpose and which appeared in the stock bottles, thirty-nine came as single individuals and only five as more than one individual."[322]

The peculiarities of Bar caught the attention of Morgan and Sturtevant. It was inconsistent with the known facts of every other gene in Drosophila. Its forward and reverse mutation rates were the reverse of those found for typical genes; it appeared to be mostly, if not exclusively, an event occurring in females; and the pattern of mutation, coming so late in oögenesis, with virtually no evidence of mosaicism, made it doubtful that mutation was the mechanism. Sturtevant and Morgan suspected crossing over. "The crossover suggestion has now been verified directly in two separate experiments. The two sex linked recessives forked (f) . . . and fused (fu) . . . lie on opposite sides of the bar locus, but are only about three units apart. . . . In the first experiment ultrabar (B^u), an allelomorph of bar . . . was

used. Zeleny has shown that homozygous ultrabar stocks give rise to reversions to wild-type with about the same frequency as do homozygous bar
stocks. In the present case females of the constitution $B/f\ B^u\ fu$ were
tested. Three reverted offspring were produced in about 6,500 flies. . . . All
three of these reverted individuals arose from crossing over between forked
and fused: one was forked not-fused and the other two were not-forked
fused. . . . In the second experiment . . . the females tested were of the
constitution $B\ fu/f\ B$. Three reversions have so far appeared—two wild-
type males and one forked-fused male. . . . It follows that reversion of bar
to normal is associated with crossing over at or near the bar locus."*[305]

Sturtevant carried on the analysis over the next two years and made two
additional discoveries. A new mutant of Bar arose which was a partial
reversion of Bar; he called this *infrabar*. Ultrabar, arising from crossing over
in homozygous Bar he renamed "double bar." There was no apparent
crossing over associated with the origin of infrabar. All of the other facts on
the frequency and the properties of the Bar reversions reported by Zeleny
were confirmed. "On the basis of these results we may formulate the
working hypothesis that both reversion and the production of double-bar
are due to *unequal crossing over*."[301]

The second observation was totally unexpected; it came from a measurement of the number of ommatidia in each compound eye of the various
Bar, infrabar, and double-bar combinations. "The most striking relation
shown . . . is that the relative position of identical genes affects their action
on facet number." Thus homozygous Bar, B/B, gave a mean of 68.1 facets;
heterozygous double Bar, BB/+, gave 45.5 facets. Similarly the compound,
B/B^i, gave 73.5 ommatidia and the heterozygote, $BB^i/+$, gave 50.5 ommatidia. Homozygous infrabar, B^i/B^i, with 292.6 ommatidia was also broader
than heterozygous double infrabar, $B^iB^i/+$, with 200.2 ommatidia.[301]

To interpret this peculiar relation, Sturtevant suggested a "hypothesis
of position effect." In this hypothesis, "it seems probable that an influence
of the relative position of genes on their effectiveness in development may
be interpreted in terms of diffusion and localized regions of activity in the
cell. This idea is, however, scarcely worth elaborating until more evidence
is obtained."[301]

Sturtevant ruled out the unlikely possibility that all mutations arose
through crossing over. By isolating spontaneous lethals from heterozygous
females carrying several linked genes, Sturtevant found that there was "no
indication the crossover chromosomes are more likely to contain new lethals
than are non-crossovers."[301] He assumed, however, that Bar was something
added to the X chromosome and had no corresponding allele. Its elimination through crossing over supported a discredited hypothesis: "It will be

---

* Not all of the revertants were crossovers and those wild types which did not show
an exchange of markers remained enigmatic until Peterson and Laughnan (1964) showed
that they were a consequence of intrachromosomal pairing of the duplicated region.

observed that the hypothesis advocated in this paper makes bar, double-bar, and round by reversion . . . represent quantitative variations of the same substance. In the case of bar and round, the hypothesis is the same as the original and most special type of quantitative view, the 'presence and absence' hypothesis."[301] Nevertheless, the peculiarities of the Bar case ruled it out as a general model for the origin of mutations. "It is clear . . . that the Bar case does not furnish support to the idea that mutations in general are quantitative in nature. Even with respect to multiple allelomorphs, where the quantitative view has often been urged, it is obvious that, at least in the cases of white and variegated, the Bar evidence does not in any way support that view."[301]

The position effect of Bar raised the possibility that it might be a general phenomenon among other genes. Sturtevant tried two dominant mutations, Delta, a wing vein abnormality, and Hairless, a bristle mutation. The number of bristles measured in the coupled and repulsed phases (i.e. the cis and trans arrangements) of these two mutants gave the same value. Position effect was a phenomenon different from the mere coupling and repulsion of factors on chromosomes.

Goldschmidt, however, considered the Bar case "conclusive proof" of his own theory of the gene. "Sturtevant . . . proved in a series of exceedingly beautiful experiments that ultrabar is the product of unequal crossing over resulting in the location of two Bar genes in one chromosome. Thus he was able to build up individuals with 1, 2, 3, and 4 such Bar genes. . . . Here then we have a series of genotypes behaving experimentally like multiple allelomorphs. But it is an experimental fact that the majority of the members of the series are formed by different quantities of a gene. . . . Thus we regard this case as conclusive proof of the quantitative nature of multiple allelomorphs."[143]

A novel interpretation of the Bar case was suggested by one of Zeleny's students, D. H. Thompson, in 1931. He assumed that the gene for the normal allele of Bar *did*, in fact, exist but that it was similar to an apo-enzyme, which he called a *protosome*, and Bar itself resembled a completed enzyme, its prosthetic group or co-enzyme designated as an *episome*. The episomes could either form chains, as in the case of double-bar or double infrabar or they could attach to different parts of the protosome, like multi-headed enzymes. The double mutant BB^i for example, when homozygous gave rise to Bar or infrabar crossovers but not round-eyed flies. Thus the two episomes were considered separately attached rather than forming a chain which could be removed by some single displacement in the crossover process. "This conception pictures the gene as consisting of a main particle firmly anchored in the chromosome with varying numbers of one or more kinds of other particles attached. The main particle is called the protosome and the attached particles the episomes. . . . Gene mutation is due most frequently to the loss of one or more episomes from the protosome and less frequently to the addition of episomes."[307] A quantitative series of multiple

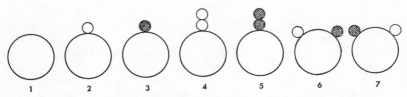

Figure 18. Thompson's Episome-Protosome Model. The protosome (large circle) was activated by an episome (small circle). The protosome without episome (1) was recessive for the Bar effect. In (2) the normal episome results in the B+ allele. A mutation of the episome (3) resulted in the infrabar allele. Duplication of the normal episome (4) gives rise to Bar. Double infrabar (5) is similar to Bar in structure. The Bar-infrabar or infrabar-Bar organization is depicted in (6) and (7). In Thompson's model, a deletion is not the loss of a segment of genetic material, but the shaving of episomes from protosomes (the concept is analogous to the term "line mutation" used by the Drosophila group when deletions were first observed).

alleles would consist of the protosome and various numbers and kinds of episomes. The peculiar observation of the (f-B) deficiency in compound with Bar could be explained by assuming that Bridges did not really have a deficiency: that segment of the chromosome merely had its episomes shaved off! Thus B/Df (f-B) was homozygous for the protosome but one of the protosomes had the Bar episome attached. This would account for the identity of the phenotypes of B/+ and B/Df (f-B), and it would render unnecessary Bridges' invocation of two sets of modifiers existing in the X chromosomes at other loci.

With the advent of radiation as a tool for inducing gene mutation and structural changes in the chromosome, a number of investigators encountered new phenomena. Muller and Altenburg in 1930, for example, had found that the number of translocations for the total genome induced by high doses of radiation was comparable to the lethal frequency (about thirteen percent) induced in the X chromosome. Most of these were detected by a suitably designed genetic scheme but on testing these, Muller uncovered a new phenomenon. "A rather surprising finding, in connection with translocations and other changes in the arrangement of the genes, was the fact that, even when all parts of the chromatin appeared to be represented in the right dosage—though abnormally arranged—the phaenotypic result was not always normal. In fact, both translocations and inversions more often than not seem to have a recessive lethal action associated with them. . . . Moreover, those re-arrangements which are not accompanied by a lethal effect commonly produce sterility and morphological abnormalities when homozygous. The reason for these effects had not been worked out satisfactorily, it being unknown whether they are due to the morphogenetic effect of the re-arrangement in itself, or to some gene mutation or gene loss ordinarily accompanying the re-arrangement. . . . In addition to these, a group of distinctly peculiar variants accompanying chromosome abnormalities has been found. These I am provisionally terming 'ever-sporting displacements.' "[232] The mottled white-eyed abnormalities were all "ever-sporting displacements" and their defect could be traced to chromosomal inversion or translocation associated with one break near the white locus. "It seems clear from this list of cases, that there is some causal connection between the type of genetic instability in the somatic cells, involving the gene at the locus of white, which gives mottled eyes, and the occurrence of a rearrangement in the linear order of the genes."[232]

Using Muller's techniques, Theodosius Dobzhansky (working in Sturtevant's laboratory) discovered a recessive allele of Bar which he called "baroid." The baroid mutation also reduced crossing over in the chromosome region near Bar and genetic tests proved it to be a translocation between the X and the second chromosome. This strongly suggested that the same mechanism suspected by Muller and Altenburg might be associated with baroid. "Muller and Altenburg suggested also that the action of a gene may be changed by 'the alteration in intergenic contiguities.' A translocation, bringing about intimate contacts between genes which were not in contact before, may change the developmental effects of these genes. This suggestion is based on Sturtevant's discovery (1925) of the 'position effect' in bar."[115] Using the position effect hypothesis with Thompson's protosome-episome chain theory of the gene, Dobzhansky tried to account for the various Bar mutations and their reversions. The "episome" in Dobzhansky's theory might be a segment of genetic material inserted on or near Bar itself. "The mutations from bar to infrabar must be due, as pointed out by Thompson, to a change in the structure of a gene located in the attached section and not in the protosome since the reversion from bar to wild-type is indistinguishable from the reversion from infrabar to wild-type. . . . The origin of baroid may . . . be due also to a position effect. The bar 'protosome' is present in all flies. It is located in the X chromosome, presumably between the genes forked and small eye. It is capable of producing the bar characteristics when subjected to the position effect of the genes located in the section attached to it in the case of bar. In baroid the X chromosome is broken very near the protosome. The normally existing association between the protosome and the genes lying to one side of it is removed. A new association with some of the second chromosome genes lying normally at vestigial is established. The effect of this new association or of the discontinuation of the old ones, is the appearance of the baroid characters."[115]

In 1932 Muller proposed a physiological classification of genes on the basis of their activities in different doses. Both Curt Stern, using bobbed, and Muller, using several other sex-linked genes on the X chromosome, observed that the addition of a third dose of a recessive gene to the homozygous mutant resulted in a normalization of the mutant effect. Using the white-eye series as a beginning, Muller studied such added doses for eosin, apricot, and white itself. "The first locus which we undertook to study was that of white eyes. We chose first flies containing the moderately pigmented mutant allelomorph of white called eosin, in which the color is considerably lighter than the normal red. . . . By irradiation we produced a deleted X chromosome containing this gene. It was then found that the addition of this fragment to a male or female which was otherwise an ordinary eosin caused the eye color to become darker, more nearly like the normal red. This shows that the actual effect of the eosin eye is not to inhibit color, as might have been thought by comparison of it with red, but to produce color, since the addition of it results in more color—only it does not produce as much color as the normal 'red' allelomorph does."[233] Muller called this type of gene a *hypomorph*.* "Tests thus far indicate that most mutant genes (both spontaneous and induced) are hypomorphs, inasmuch as they show 'exaggeration' with deficiencies as Mohr has pointed out, or at least give a form having about the same degree of abnormality as the homozygous mutant. The latter relation would be expected in cases like white eyes, where the mutant gene had nearly reached the bottom of the scale of effectiveness and hence itself has almost as little normal effect as the deficiency had. This latter type of mutant may, descriptively, be called 'amorphic.' "[233]

If three doses of a hypomorphic gene were more normal than two doses of the same gene, why then were males and females alike for most of their genes on the X chromosome when the male had only one dose and the female had two? Muller, like Stern, suggested that a mechanism of "dosage compensation" had evolved to equalize the effects of the sex chromosomes in the two sexes. "This must be, of course, due to the interaction of other genes in the X chromosome, whose simultaneous change in dosage affects the reaction. In some cases, at least, it has been possible to show by studies of the effects of different chromosome pieces, a) that genes other than genes for sex are acting as the 'modifiers' in question, b) that the modifiers responsible for the dosage compensating effect on different loci are to some extent different from one another, and c) that more than one modifier may be concerned for a specific locus."[233]

Thus the studies on doses of genes in the X chromosome provided two new characteristics of the gene concept—a physiological interpretation of the direction of action of the mutant gene and a theory of modifiers acting on the X chromosome genes to abolish the dosage effects of the normal genes in the female.

Not all mutations could be interpreted by hypomorphic and amorphic activity of genes. Some acted in a direction antagonistic to the normal; they could be considered inhibitors of normal activity. Extra doses of such genes tended to exaggerate the mutant character. These inhibitor activities Muller called *antimorphic*. Another type of activity exaggerated or increased the normal activity of genes; most reverse mutations would be examples of such *hypermorphic* activity. The most intriguing type of activity, however, was

---

* The microbial geneticist today uses the term "leaky mutant" for hypomorph and "non-leaky mutant" for amorph.

that which had no relation to the normal activity of the gene from which it arose. In this *neomorphic* class, Muller placed Bar eyes. The normal allele of Bar was amorphic with respect to Bar. "The fact that normal genes may thus act as amorphs with regard to a particular character affected by their mutations should serve as another warning against regarding mutant genes that seem to be amorphic or hypomorphic as really involving a mere absence or loss of material."[233]

Bar was clearly unaffected by its normal allele because the Bar deficiency did not alter the phenotype of Bar in compound with it. "Bridges' original Bar deficiency of 1915 (published upon in 1917), which we may now interpret definitely as a loss, shows that the absence of the Bar-locus in the non-Bar chromosome of a heterozygous Bar female has the same effect on the Bar eye character as the presence of the normal allelomorph itself. . . ."[233] Rather than thinking of Bar as a "new presence" Muller considered it to be a neomorphic mutation, or because of its peculiar properties in reversion to round eyes, "there is a possibility that in the origination of Bar a gene became duplicated *in situ*, and that one of the resulting twins mutated at the same time. . . ."[233]

The possibility that Bar was such a duplicated gene could not be checked cytologically in 1932, but a few years later a new technique was introduced to genetics. In the salivary glands of larvae of *Drosophila* and other Diptera, the nuclei contained peculiar chromosomes which, when spread out, were about one hundred times bigger than the best mitotic preparations available from dividing cells. These "giant chromosomes" discovered by Heitz and Bauer led T. S. Painter to suggest their use for cytological studies of chromosomal abnormalities in *Drosophila*. Virtually every *Drosophila* geneticist in the mid-1930's tried his hand at this technique. The most successful application was made by Bridges, who had developed remarkable cytological skills through Wilson's influence and association. In the Soviet Union, Muller enlisted the services of A. A. Prokofeyeva. Her cytological skills, while not as spectacular as Bridges's, developed to a point where Muller felt a search could be made for the Bar "duplication." This was found in 1935 and published early in 1936 by Muller, Prokofeyeva, and K. V. Kossikov with the title "Unequal Crossing Over in the Bar Mutant as a Result of Duplication of a Minute Chromosome Section." The segments in the Bar duplication "follow in the same order, not mirror wise. This condition obviously arose in the first place by means of breakage and mutual translocation between homologous or sister chromosomes at nearby but not quite the same point. . . . The Bar phenotype . . . represents the 'position effect' of the extra section of chromatin, interacting with another locus or loci nearby in the chromosome."[250] Sturtevant had argued that Bar did not have a normal allele corresponding to it and Demerec supported this view because numerous round-eyed reversions appeared from x-rayed Bar males, suggesting that the Bar gene had been deleted by the action of the x-rays. It also suggested that lethal mutations themselves may arise

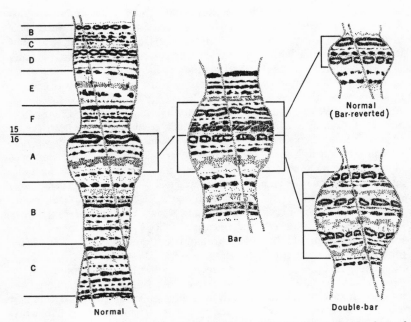

B
C
D
E
F
15
16
A
B
C
Normal

Bar

Normal
(Bar-reverted)

Double-bar

Figure 19. Bar Eyes—a Duplication. Bridges and Muller independently observed that Bar was a cytological duplication visible in salivary chromosomes. The figure, from Bridges's analysis, indicates the doubleness of Bar and the tripleness of double-bar for the region 16A. Bar-reverted, which arises by recombination from homozygous Bar stocks, shows the normal, unduplicated, 16A region.

chiefly from the deletions of their normal alleles. This view was no longer satisfactory if Bar were a *bona fide* duplication. "It is . . . fallacious to consider the 'reverse mutation' of Bar to normal as constituting deficiency of an individual gene as Demerec had done, and his inferences concerning the viability or inviability of single gene deficiencies in general thus lose validity, since the Bar analogy was used in inferring that only single genes were lost in the other cases."[250]

Independently of Muller's analysis of the Bar region, Bridges that same year presented a paper on the same topic, "The Bar 'Gene,' a Duplication." "A chance to clear up some of the puzzles as to the origin and behavior of Bar was offered by the salivary chromosomes. Study of the banding in a stock of Bar . . . showed that an extra, short section of bands is present in excess of the normal complement, forming a duplication."[54] Both normal and reverted-Bar chromosomes showed the unduplicated segment, Bar itself showed the duplication, and double Bar was a triplication! Bridges, however, interpreted two aspects of the Bar case in a way that differed from Muller's interpretation. Bridges believed that the Bar region of one chromosome had been deleted and inserted into a break near the Bar region of the homologous chromosome. Also, he believed there was no difference between a quantitative interpretation and a position effect interpretation for the repeated segments of the various alleles. "The Bar-eye reduction is thus seen to be interpretable as the effect of increasing the action of certain genes by doubling or tripling their number—a genic balance effect. But 'position effects' are never excluded when duplications or other rearrangements are present, either in the wedging further apart of genes normally closer, or by the interaction with new neighbors. In Bar and its derivatives the respective shares attributable in the total effects to the genic balance change and to the position-effect change seem at present a matter of taste, rather than of completed analysis."[54]

Muller objected to these two points. "We would take issue . . . with Bridges' designation of the duplication as an 'inserted' piece, the point of 'insertion' of which (to right or left of the original single section) is uncertain. . . . The more probable explanation for this case . . . is that the two sister or homologous X chromosomes were already separate at the time of breakage, that only the left hand break occurred in one of them, and only the right hand break in the other, and that in the subsequent process of attachment, the left hand piece became attached to the right hand piece of the chromosome having the breakage further to the left. On this more probable view, then, the duplication did not originate as an insertion at all, and the original mutual breakage point is the point where one twin section now joins the other one. . . . It is evident that, on this view, the Bar duplication has originated by a process which may be termed 'unequal crossing over.' The unequal crossing over ordinarily observed in Bar is, then, only a kind of secondary unequal crossing over, resulting indirectly from the primary unequal crossing over which established the duplication in the first place."[*][238]

The differences between a position effect and a quantitative balance interpretation of the abnormalities seemed more striking to Muller than to Bridges. "We would also take issue with Bridges's opinion that the phenotypic effect, Bar eye, may, according to 'taste,' be considered either as a result of a relative dosage change of genes in the duplicated section, or as a 'position effect.' For evidence has been found . . . that Bar behaves as a neomorph. That is, the mere addition of extra doses of the general region in question . . . does not increase the Bar effect. On the other hand, other rearrangements of genes in the Bar region . . . do cause effects similar to those of the Bar duplication. . . . The phenotypic change is therefore solely a result of the 'position effect,' and this effect must be sharply distinguished

---

* Muller's choice of the terms "primary and secondary unequal crossing over" in this criticism suggests a deliberate "one-upmanship" based on the parallel terms "primary and secondary non-disjunction" which Bridges used in his Ph.D. thesis.

from the effect of dosage change, even though in many individual cases in genetics it has not yet been possible to judge with which class of effect we are dealing."[238]

It is ironic that the main evidence for Bar as a neomorph was derived from Bridges's own observations almost twenty years earlier that the deficiency for Bar had no effect on the broadened eye of the B/Df(f-B) female. The real significance of the Bar case, however, went far beyond the "academic" controversies on its origination, function, and structure. "We consider the point of chief interest in the Bar case to be its illustration of the manner of origination of extra genes in evolution. Bar had for a long time offered the best case yet known for the idea that genes could arise de novo. Its interpretation as some sort of duplication met with difficulties, in our ignorance of the real existence of a 'position effect' on non-allelomorphic genes upon one another. Now these difficulties are resolved and there remains no reason to doubt the application of the dictum 'all life from pre-existing life' and 'every cell from a pre-existing cell' to the gene: 'every gene from a pre-existing gene.' We need at present make an exception here only of those very special conditions under which life itself, as a naked gene, originates."[238]

CHAPTER 14

Position Effect: Lethality, Variegation, and Mutation

To the beginning biology student who has never pursued genetics beyond the half dozen or so lectures he encountered as a college freshman or sophomore, the story of genetics begins with the rediscovery of Mendel, achieves classical status with Morgan, and rapidly moves to modern times with microbiology and nucleic acid chemistry contributing most to his knowledge of the gene. Even in a genetics course it is not uncommon to find the period beginning with the year 1927, with the induction of mutations through radiation, quickly spanned to the 1940's, with the first breakthroughs in biochemical genetics. The 1930's, even in relatively advanced courses in genetics are frequently passed by as a period of genetic "Depression." Yet the 1930's represent an exciting period of time in the evolution of the gene concept. They might be considered the zenith of sophistication, intricacy, and mastery of cytogenetic techniques for the exploration of the gene concept in *Drosophila*. In this same period, too, maize genetics offered the first substantial challenge to *Drosophila* as the tool for studying the gene concept.

Throughout this decade one tool—radiation—emerges dominant to all others in working out the various challenges and complexities faced by the gene concept. Between 1927 and 1930 the Coolidge tube proved to be a cornucopia from which the riches of lethal mutations, recessive visible mutations, ever-sporting displacements, gross structural rearrangements and minute structural rearrangements poured in generous quantities. This profusion of discoveries could be classified in a more simple way. "It is . . . clear

117

that these heritable changes are of two main kinds: changes in what I may
call the gross morphology of the chromosomes, and ultra-microscopic
changes of the sort that have been termed 'point mutations' or 'gene
mutations.' "[231] Many geneticists, however, did not share this view and
looked upon the radiation mutations as destructive changes rather than
changes similar to those arising naturally. Their utility for evolution was
questioned and their significance for exploring the gene concept was considered negligible or misleading. To this battery of criticisms, Muller offered
a defense of "Radiation and Genetics" in 1930. "Can it be that we have in
the X-ray tube and in the radium needle merely an amusing but not very
instructive toy, wherewith to produce all sorts of bizarre monstrosities, that
are very pretty for us to play with, but that, after all, these 'laboratory
deformities' can have but little bearing on the constructive evolutionary
processes of organic nature and on their physical basis? It has been hinted
in more than one place that all we have done is to break and knock holes in
the chromosomes; perhaps, too, to get the pieces tangled together sometimes, but that such stunts can no more help us in a real understanding of
one object than would the wrecking of a train help us to understand the
normal workings of a locomotive. Are we simply wreckers? How can we
meet such criticism?"[231]

Muller's defense certainly excluded some poorly defined criticisms and
oversimplifications about the radiation effects. The differences in the ratios
of "point mutations" to gross rearrangements for male and female germ cells
made it unlikely that the agent was acting in the same way on both of them.
Reversions of x-ray induced forked bristle mutations made it seem unlikely
that the "holes" punched in the chromosome involved complete genes and
equally improbable that extensive losses of parts of a gene could have
accounted for the original induced forward mutations to forked. As was
found for spontaneous mutations, "so far as I have been able to ascertain,
there is no greater proportion of lethals, as compared with visible mutations
among the X-ray point mutations."[231] Also, the spectrum of multiple alleles
at the white locus could be duplicated with x-rays, providing many whites
as well as apricot, eosin, and tinged alleles. All of these findings gave some
hope for maintaining that radiation was not a "wrecking" agent but that
its effects, whether direct or indirect, might be worked out. The scope of
radiation genetics was vast; with each experiment the problems increased
faster than the answers. Some of these problems were to lead into developmental genetics, some into the cytogenetic analysis of gross structural
changes in the chromosomes, some to physiological studies of dosage effects
and dosage compensation, and some to the mechanisms of mutagenesis. But
the foremost field in Muller's mind was the gene itself. "Another field that
we are only at the edge of is that of gene study: for example, determining
what the gene in given loci are capable of if we cause them to mutate again
and again, both in parallel (i.e., when the same allelomorph mutates on
different occasions to various other allelomorphs) and in series (when a

given gene undergoes successive changes from one allelomorph to another); counting and maybe measuring genes; studying their stabilities; their dominance relationships; their effects in different relative quantities (deficiency and excess); their influence upon the synaptic and other properties of the chromosome or chromosome region containing them; their possible position-effects if they are shifted in location."[231]

The simpler problems came first. The rearrangements of the chromosomes easiest to detect by use of a genetic design which would give quantitative results, were the translocations. The first translocation, found in 1918 by Bridges, "stood alone, however, as a genetic oddity, unparalleled by similar cases that could be satisfactorily proved, until 1926."[247] Curt Stern had found an additional translocation that year, demonstrable both cytologically and genetically. But in the following two years an abundance of translocations was found by Muller and Altenburg. "In all there were 21 translocations found among the 161 F_1 flies (representing as many X-rayed sperm cells of P_1 tested), or 13 percent. This was a startling high frequency, in view of the previous uniqueness of the phenomenon, and was comparable with the rate of detectable gene mutations occurring after X-raying."[247] Two facts were immediately apparent from an analysis of the translocations. The points of breakage among the various translocations were not distributed uniformly along the mitotic length of the chromosomes; and the "majority of the translocations were lethal when homozygous."[247]

Although the possibility that simultaneous breakage and mutation at a nearby gene could not be ruled out, Muller favored a second hypothesis. "The alteration in intermolecular surroundings of the genes directly adjacent to the points of breakage and reattachment, in other words the alteration in intergenic contiguities, has in itself brought about a change in the quantity or quality of the physico-chemical action of these genes upon the protoplasm, so as to make them, in effect, somewhat different genes, as though gene mutations had taken place in the genes on either side of the breakage and attachment points. Plausibility is lent to such an assumption through Sturtevant's finding that two genes for Bar eye adjacent to one another in the same chromosome seem to have an amount of effect on the development of the eye different from that of two otherwise identical genes for Bar eyes that lie in separate, homologous chromosomes."[247] Position effect, as the basis for the lethality of most of these homozygous translocations, was a distinct possibility.

The difference between the paucity of natural translocations and their unexpected abundance from high doses of radiation was a real one, however, whose mechanism would not be resolved for several years, when quantitative studies of translocations with respect to different dosages of radiation would become more refined.

Among the many discoveries coming from the widespread use of radiation in genetic research in the first few years after its introduction were two that gave an insight into the position effect of genes associated with chromosomal arrangement. The first was from Muller's initial experiments. When Muller reported the discovery of "eversporting displacements" he was careful not to choose the term "eversporting genes" because the mottled white which arose showing this variegation was also associated with a similar variegation at the nearby Notch wing locus. The mottled-eyed stock quickly produced two varieties, a dark-eyed strain with occasional flecks and a light-eyed strain with red flecks. "The lights consistently average much more extreme Notch than the darks which often show no Notch at all; the light progeny so produced tend to transmit their more extreme Notch to their own light mottled offspring, whereas the dark progeny tend to segregate again into two major classes (dark, less Notch and light, more Notch) in the following generation. This makes the Notch character 'eversporting' in the same way, and in parallel with the mottled character, and indicates that the phenomenon does not depend on an assortment of individual 'gene elements' or of any particles within and smaller than a single gene, but rather depends somehow upon the peculiar behaviour of chromosomes or segments of chromosomes, larger than and containing more than one gene."[232]

Additionally, "the break was at some distance to the right of Notch, and there are some unmutated genes in the right portion of the fragment, between Notch and the break, and others in the left portion between the left end of the chromosome and mottled."[232] From these two observations, later to be designated as "spreading effects" and "skipping effects," Muller argued that "eversporting gene" was inappropriate and that the phenomenon was more aptly an "eversporting displacement."

The other discovery was made by J. W. Gowen and E. H. Gay in 1932. In a study of white mottled eyes, they found a similar pattern of light and dark types, the latter segregating into the two forms as Muller had found. But a more detailed analysis of these red-eyed types revealed a startling discovery. "Breeding red eyed flies reveals that they fall into two classes: one ... normal overlaps which breed exactly like mottled flies; the other comparable with Muller's dark mottled class—a different genotype. This latter type of eversporting we have attributed to the Y-chromosomes—a phenomenon hitherto undescribed."[155] The presence of an extra Y chromosome in either the mottled female or mottled male would repress the variegation, rendering it normal or nearly normal. The following year Gowen and Gay reported that temperature also affected the expression of mottled eyes, cool (18° C) temperatures increased the mutant characteristics of the mottled and lowered their viability while at warmer (24° C) temperatures the eyes were partially normalized and the viability of the mottleds was significantly higher. This type of variegated position effect was unstable. In 1936 J. Schultz demonstrated another unusual feature of these variegated position effects, as he called them; they all involved the juxtaposition, next to the variegation, of a region of the chromosome known as *heterochromatin*. The heterochromatin was usually located near the tips and centromeres of chromosomes. The variegated genes, on the other hand were located in the euchromatic portions of the chromosome adjoining the heterochromatin brought in by the rearrangement. Not all of the translocations were position effects of this sort and the addition of an extra Y chromosome (which is almost entirely heterochromatic) had no effect on these or on such purely euchromatic position effects such as Bar eyes.

In 1933 Muller left for the U.S.S.R. where he attempted to develop formal genetics on a large scale. Active groups of geneticists working with Drosophila were soon established in Leningrad and Moscow, and their influence soon spread to other major research institutes in the U.S.S.R. Some of these groups studied gene structure, some studied position effect, and others studied the quantitative laws of radiation mutagenesis. As a consequence of these studies, a unique interpretation of position effect was proposed by C. A. Offermann in 1935. In his paper "The Position Effect and Its Bearing on Genetics," Offermann proposed a classification of genes based on their capacity for responding to position effect. Some genes were potentially sensitive to change if their immediate neighboring genes were changed or removed. Other genes did not show this sensitivity or, when altered, were capable of exerting such a position effect on one or more of their neighbors. "Thus if a mutates and this mutation (like a loss of a from the neighborhood of b through rearrangement) exercises a position effect on b, then the expression of b will be changed. Since, however, b itself did not in this case change its own structure, it can be replaced by another normal b, without the expression being further altered; the same relation does not, however, hold for a. In such a case a test by crossing over will give us a as the location of the mutation, whereas a test by the use of deleted chromosomes to cover or uncover the locus will give us b as the location. If a gene employed as a reference in both tests should lie between the two participants, the results on the gene order obtained by the crossing over and by the deletion methods would seem contradictory."[225] As many as three positions on the genetic map could be assigned to one mutation that was subject to position effect: a "true" mutation at the sensitive locus itself; a mutation of one of its neighbors to the immediate left which, in itself, would not express a mutant phenotype, but which would exert a position effect on the sensitive gene, making it appear as if it had mutated; and a mutation in one of the genes to its immediate right, which would also exert a position effect without necessarily exhibiting any mutant effect of its own. The extraction, by crossing over, of the unmutated "central" gene in this situation would have led to the mistaken impression that the site to the left (or the site to the right) of the sensitive gene was the actual locus for its position on the map, when in fact the gene was at neither one of those two sites! Offermann's analogy was based on the finding, two years before, by Dubinin and Sidorov that the variegated gene near the point of breakage in a chromosome rearrangement could be replaced by its allelic point mutation through crossing over, and in such cases the extracted "variegated"

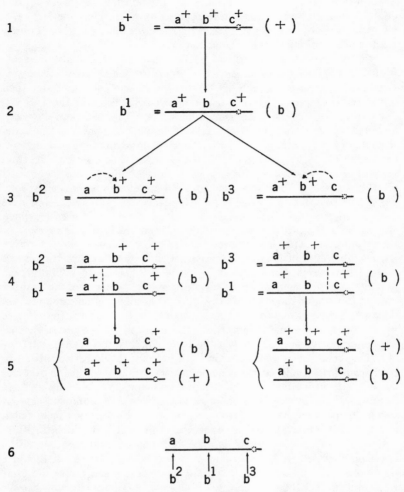

Figure 20. Offermann's Paradox. In 1, a normal gene b^+ is bounded by genes a^+ and c^+. In 2, mutation of b^+ to b by simple gene mutation results in a mutant (b) phenotype. In 3, a mutation of a^+ to a exerts a position effect on b^+ leading to a (b) phenotype. Similarly, mutation of c^+ to c exerts a position effect on b^+. In 4, two alleles of b^1 are tested for recombination, and in 5, both yield crossovers. If appropriate marker genes are also used, the alleles may be mapped in the order shown in 6. Two of these positions for b would occupy the sites of genes functionally distinct from b.

gene would be perfectly normal. Similarly, the normal gene could be reintro-duced by crossing over with the apparently non-variegated allelic amorph at the site of the chromosome break; the normal gene, after crossing over, would assume the variegated expression.

All of these revelations clearly showed that position effect had no per-manent effect on the gene itself. But if position effect could simulate true gene mutations, as in Offermann's hypothetical model, would it not be pos-sible that all mutations were due to position effects? Offermann did not think so, agreeing with Muller that the course of evolution demanded some sort of qualitative changes in the genes themselves. With the advances in salivary chromosome analysis, Muller and his colleagues were able to demonstrate the existence cytologically as well as genetically of minute chromosomal deficiencies and inversions. Offermann's question was no longer academic. "The general question thus arises, what proportion of apparent mutations are only intergenic 'position effects' rather than autonomous intragenic changes? Of twenty-seven scute and achaete mutations investi-gated which have been produced by irradiation, it has so far been possible to demonstrate in eighteen cases that there was a breakage and re-attach-ment close to the scute or achaete locus. Some or all of the remainder also are probably intergenic rearrangements, for it has been found in this investi-gation that the rearrangements tend to fall into two categories, gross and minute, the latter being of such a nature that a genetic discrimination between them and true intragenic mutations would be very difficult, or in many cases even impossible. . . . This question is of moment because the range of possibilities of phenotypic change through intergenic rearrange-ment alone must be far from adequate for any indefinitely continued evolution."[251]

Two conceptual limitations emerged from the attempt to define point mutation. On the one hand, there was an absence of a positive technique for demonstrating point mutations to be chemical modifications of the gene; on the other hand, there was an ambiguous classification of point mutations as the residuum of all mutations that were neither demonstrably gross nor minute rearrangements. The limitations appeared irreconcilable and it was no longer possible to avoid a confrontation with his own alterna-tive hypothesis. The stage was now set for the conflict between the advo-cates of gene mutation and the advocates of rearrangement by breakage, each side believing that its hypothesis could provide the basis for the muta-tions significant for evolution as well as for those mutations relegated to the "residuum" by cytogenetic analysis.

CHAPTER 15

Position Effect: An Attack on the Gene

For a period of twenty years Richard Goldschmidt argued for a compound gene, consisting not of two types of "gene elements" as Eyster had proposed, but of only one element—the gene itself. Multiple allelic series were proportional to the quantity of the gene present in the cells during their various developmental and physiological activities. "I have never been able to understand why this conclusion, which safely rests on experimental facts, has been considered by some as offensive. The number, the size, and the shape of the chromosomes are constant; the size of the cells is constant and often their number in a given organ; the number and size and arrangement of blastomeres are constant, the number of segments, of bristles, and I know not what else. Orderly development of a given organism requires a wonderful amount of quantitative constancy from the organ down to the chromosome. Why then should exactly that bit of substance which after all is responsible for all the rest be required to produce its wonderfully typical action of a unique sameness on the basis of a negligible quantity?"[144] For the geneticist raised on the chromosomal mechanics and mapping procedures of *Drosophila*, Goldschmidt's physiological interpretation was difficult to follow. "Expressed more specifically, the genes must be things which produce their typical effects by catalyzing chains of reaction, the speed of which, ceteris paribus, and given the specific substance of each gene and the plasmatic substratum, is proportional to the quantity of the gene and therefore fixed within the entire system of simultaneous coördinated reactions of different speeds."[144] This attempt to define

124

the gene on the basis of its inferred physiological action lacked the experimental productivity of the mutational and crossover analysis of the genes themselves. Goldschmidt thought that the Bar case gave him a demonstration of quantitative multiple alleles and an explanation of position effect itself, but the cytological studies of Muller and of Bridges, as interpreted by Muller, ruled out such a hypothesis. The essential difference between Goldschmidt's point of view and the concept of the "classical" gene was expressed by Goldschmidt on numerous occasions as a difference between "dynamic" and "static" outlooks. "It is of no use to discuss the problem of the quantity of the gene without considering the corresponding reaction velocities through which alone the assumption of different but typical gene-quantities becomes important; because without this connection we have only a sterile hypothesis."[144]

Goldschmidt dropped his previous theories of the gene when position effect was found to extend to numerous gene rearrangements from gross to minute alterations. In 1937 Goldschmidt argued that "the time has come to acknowledge that gene mutations have as little existence as genes themselves. (A number of geneticists have already played with this idea, but hesitated to drop the old conception of the gene. They took refuge in position effects and mutations near the locus of a break.) The idea of a position effect, made to save the gene concept, will also have to disappear when it is recognized that the position effect is actually identical with what was called a gene. The chromosome as a unit will be found to control normal development or wild type. The changes of the correct order within its chain produce deviations from normal development, called mutants. Though they are localized, there is no such thing as a gene and certainly no wild type allelomorph."[146]

Goldschmidt's evidence, other than the abundant cases of position effect from Muller's own radiation experiments, included several "poly-mutant" changes in some lines of Drosophila which he studied. Goldschmidt considered these as instances of massive chromosomal rearrangement which so upset the normal chromosomal organization that multiple abnormalities were expressed. In this respect, he rejected the factorial theory of Darwinism which was based on the mutation of the individual gene and proposed instead that "macromutations," similar to the jumps originally advocated by de Vries, were more significant for evolution. The former difficulties with the Bar locus could now be resolved through a simple position effect interpretation. "Anything happening at the Bar locus may produce a Bar effect. This change then acts in the language of classic genetics like a multiple allele of the respective mutant. It is obvious that it is very difficult in such a case to distinguish between a position effect and a gene mutation. The only difference is that in one case we know that the phenotypic effect is produced by a rearrangement of chromosomal material; in the other case we do not know this yet. The suspicion is strong that the gene mutation at the locus in question is also the result of a rearrangement, maybe one which is difficult to make visible cytologically on account of its involving only a very small chromosome section. . . . The conclusion, then, is that gene mutation and position effect are one and the same thing. This means that no genes are existing but only points, loci, in a chromosome which have to be arranged in a proper order or pattern to control normal development. Any change in this order may change some detail of development, and this is what we call a mutation. We might of course call a change of arrangement at a locus, a gene. But then there are no genes in the normal chromosome, and the mutant gene has no wild type allele, as the whole wild type chromosome is the allele for all mutant genes in the chromosome. Better then, give up the conception of the gene except for simple descriptive purposes."[147]

To account for this unique wild-type pattern in a continuum, the chromosome in Goldschmidt's model was a single organic molecule, similar to that proposed by Castle in 1919 in his model for recombination. The prevailing view that such a molecule would be a protein was not an essential feature of his model; this prevailing view had no basis for changing until bacterial transformation was shown to be associated with nucleic acid. "Let us assume that the individual chromosome actually is a single immense chain molecule and a proteinase (this means the essential part of the chromosome—the so-called gene-string—to which is added nucleic acid, which makes in some way the stability of the long chain possible. . . .) This proteinase then has different properties according to its special surroundings. Either it may produce its own replica, which amounts to a division of the chromosome or it may synthesize other proteins from the parts present or it may hydrolyze proteins."[147] In contrast to this unitary hypothesis of the chromosome, Goldschmidt attributed an outmoded and old-fashioned quality to the gene concept itself. "It is my opinion that the classical theory of the gene corresponds to the conception of the individual atom of old physics. Genetics has outgrown this, I hold, and finds itself in a condition parallel to that of physics immediately before Rutherford. I am sure that our Rutherford stage will soon be reached, and then we shall be ready for our Planck and Bohr. I am sure that their model of the hereditary material will be the model of the chromosome and not of the gene."[148]

The strongest argument against the classical gene, however, was the unexplained phenomenon of position effect itself. "In Muller's language this showed that the normal gene, affected in its function by the neighboring rearrangement, returned to its normal function when switched over to the normal chromosome, and that the normal gene, now brought over into the neighborhood of the rearrangement, became changed in its function. I cannot help seeing serious logical difficulties in this interpretation. If a real mutation is a chemical change within a gene considered to be a molecule, we are required to believe that the completely unchanged gene produces just the same visible effect as the changed one if the linear order or structure is interrupted in its neighborhood. Nothing has changed chemically within the gene or without, all the 'genes' are there as before, only their order has been reshuffled, and nevertheless the same effect occurs as after an irreversible change."[149]

Despite these strongly worded attacks on the gene concept, Goldschmidt did not have a strong body of experimental evidence for his own theory. His previous record of accommodating all apparent attacks on the gene—Castle's contamination theory, the use of the Bar case to justify a quantitative theory and, when that proved to be invalid, to prove his position effect theory—all tended to discredit the strength of his arguments. For the most part, Muller and other supporters of the gene concept ignored him. This was a marked contrast to the attitude of the Drosophila group some twenty years before. Of course, Castle had an experimental system which, on face value, appeared to be subject to allelic contamination and genic fluctuation. Bateson had a system with high predictive value for many of his coupling crosses. In the presence and absence hypothesis, the apparent loss of a material basis was a reasonable assumption to make for a loss of genetic function. Goldschmidt's experimental evidence was more diffuse. The unanalyzed types of rare "polymutant" events which he encountered lacked the tangible and immediate reality of the cytological examination of salivary chromosomes or the mechanistic attractiveness of the chromosome maps constructed by crossing over. By contrast, the demonstrated capacity of a small fragment of the X chromosome ("half" of a salivary band) to function normally for three of the four genes it carried when it was inserted into another chromosome was an event that seemed easier for Muller and his colleagues to interpret on a genic hypothesis than on a "pattern" of chromosomal behavior.

Goldschmidt, however, persisted. His analogies became elegant. "If I stop the A string of a violin about an inch from the base, the tone C is produced by the string. This does not mean that the string has a $+^{c}$ body at the point which, when stopped becomes C."[150] While elegant, they were nevertheless partisan, since the "violin" model he proposed could just as readily have been a discontinuous "piano" model if he had argued for the "classical" point of view! Some of his arguments were enhanced by the absence of the critical information that would have immediately invalidated them. "One of the great difficulties of the theory of the gene is the fact that the mutant locus again reproduces its kind faithfully and that therefore the assumed chemical change within the gene does not change the ability of the molecule to synthesize its own copy. This ability practically restricts the possibility of mutation to stereoisomeric changes, or surface changes according to the new viewpoints."[150] Such an argument about the structure of nucleic acids, from a contemporary point of view, is wrong; apparently Goldschmidt's "straw man" interpretation of the "classical" gene was a three-dimensional, probably globular, molecule. But Muller in 1922 had argued for its linearity, a viewpoint that he endorsed again in 1935 after salivary chromosomes permitted a new method for calculating gene size and shape. Also, Muller had argued against associating the gene with any given chemical component of the cell in 1921 and 1922 because there was no evidence to favor even the probable proteinaceous basis for it. The reproductive capacity that was retained after mutation, Muller had singled out as the strongest argument for the uniqueness of the genetic material—no other material, except viruses, had such "convariant reproduction." When it was shown that tobacco mosaic virus could be crystallized and its composition was shown to be more than 90 percent protein, Muller considered this "virtual" proof that the gene itself would prove to be protein, but a protein with that same special replicating capacity.

Although modern genetics had begun to associate a material reality—nucleic acid—for the genetic material, and substantial theories of gene action were proposed during the 1940's, Goldschmidt did not abandon the main feature of his argument—the genetic continuum of the chromosome and its pattern modifications through chromosomal rearrangements. In 1954, in one of his strongest statements on the subject, "Different Philosophies of Genetics," Goldschmidt pointed out the underlying assumptions that he and his critics used in arriving at their interpretations of the genetic material. It is a significant document because the critical experiment for exploring the problem was to come one year later. "There is no historically recognized science of theoretical genetics comparable to theoretical physics or natural philosophy, as it is called in England. But each thinking geneticist, in interpreting his factual data and in trying to fit his results into the total theoretical structure of his science, does it under the conscious or subconscious influence of his basic philosophy or *Weltanschauung* in regard to genetical thought. I mean to say by this that, when we interrupt our experiments to do some constructive thinking, we are likely to draw frequently widely divergent conclusions from the same facts. It is not that the facts are ambiguous or insufficiently established; it is the way we are looking at facts that is different."[153]

One way of looking at the facts, the "statistical" approach, is reminiscent of Castle's earlier arguments concerning the "factorial" hypothesis. "The statistical basic philosophy tries to interpret every generalized set of facts by the introduction of more and more units for statistical treatment. . . . Statistical thinking tries to explain all basic features of genetic phenomena by introducing more genes in the form of modifier systems built up by selection. In this way, a system is finally established, which is so conspicuous in much of present day genetics, and which I must call hyperatomism and hyperselectionism."[153] In contrast to this, Goldschmidt offered his physiological approach. "Although . . . the physiological, or dynamic, approach . . . accepts, naturally, the basically statistical tenets of genetics, it tries, actually within the rule of parsimony, to avoid looking for explanations in terms of unproved, additional systems of units for more and more genic permutations."[153] Once again Occam's razor proved a temptation for the geneticist. "The great difference between the two interpretations of the same facts is that the statisical one leaves room only for more and more genes of the same type and has to invent specific features, such as position alleles, to explain facts beyond the scope of the classical gene. The dynamic interpretation . . . unifies the different facts as the product of one structural principle. . . . If sections of basic action are integrated into the next hierarchical unit, one can conceive—forgetting completely the gene—that they control, again as a unit, nearly related but branching chains of reaction, which control larger elements of the developmental process. I would look for such an explanation if it is found that mutants affecting the same organ are frequently located within a larger section of the same chromosome."[153]

The "emergent" quality of Goldschmidt's theory was extended to all of life itself and to many, if not all, aspects of the physical world. "The hierarchical order is clearly essential in living nature, although it exists in inanimate nature, as the order nucleus and electron, atom radical, molecule, macromolecule, crystal shows, each higher member of the hierarchy being composed of the lower ones but different in its qualities from a mere sum of these. It is clearly not the sum but the orderly relationships of the components that are responsible for the actions at the different levels of the hierarchy. Therefore, at these different levels also, new types of interrelationships appear, say in inorganic nature the van der Waals forces on top of ordinary valencies. In view of such facts a biologist, studying a clearly hierarchical system of activities like that of chromomere, chromosomal segment, chromosome, genome, would hardly expect to meet with a situation even simpler than that present in inorganic nature—namely, total action being the sum of all partial actions, as assumed in the classical theory of the gene. . . . I am convinced that the new way of looking at the nature of the genetic material will have to supplant the statistical classical theory of the gene, before the attack on the ultimate biochemical problems is possible."[153]

Goldschmidt maintained his "continuum" model until his death in 1959, some twenty years later. Unlike Castle, he never had to face the ordeal of a public retraction because Castle based his opposition to the "factorial" hypothesis on a series of experiments which were readily adaptable to further tests, including the critical ones which proved the "residual" heredity which he had fought against for so many years. Muller (and for that matter, any other *Drosophila* geneticist) could not do more than provide indirect "proofs" for the existence of gene mutations. The foremost of these arguments was the necessity of gene mutation for evolution. A second argument was the spontaneous revertibility of mutations. How could multibreak rearrangements revert spontaneously? But neither was a direct demonstration, like non-disjunction, of a theoretical interpretation. On the other hand, Goldschmidt could not rule out chemical change in the gene. His theory never attempted to explain the apparent discontinuities in the chromosome reported by Muller and his colleagues in the analysis of the genes at the tip of the X chromosome. It was the lack of a *critical* experiment for either point of view that permitted Muller and many others to ignore Gold-schmidt's attack as "philosophical." Other critics did not enjoy this silence and they encountered swift criticism when specific experimental procedures could be brought into a debate.

Goldschmidt's outlook, however, had one important influence on the development of the gene concept. His "continuum" model permitted a climate of acceptance, rather than incredulity, when attempts were made to demonstrate that the gene itself might be a linear continuum.

CHAPTER 16

Point Mutation

In 1927, before interest in position effect extended beyond the Bar case and before radiation was widely used by other geneticists, Muller confidently claimed that "it has been found quite conclusively that treatment of the sperm with relatively heavy doses of X-rays induces the occurrence of true 'gene mutations' in a high proportion of the treated germ cells."[228] Muller based his claim on the striking parallel between the types of mutations induced by radiation and those that he had obtained spontaneously or through slightly increased temperatures. "All in all, then, there can be no doubt that many, at least, of the changes produced by X-rays are of just the same kind as the 'gene mutations' which are obtained with so much greater rarity, without such treatment, and which we believe furnish the building blocks of evolution."[228] In this optimistic spirit, Muller suggested an immediate exploration of the gene problem. "It appears that the rate of gene mutation after X-ray treatment is high enough, in proportion to the total number of genes, so that it will be practicable to study it even in the case of individual loci, in an attack on problems of allelomorphism, etc."[228] This, and other approaches which he suggested, constituted a strategy for the long range exploitation of radiation genetics. "From the standpoint of biological theory, the chief interest in the present experiments lies in their bearing on the composition and behavior of chromosomes and genes. Thus we hope that problems of the composition and behavior of the gene can shortly be approached from various new angles, and new handles found for their investigation, so that it will be legitimate to speak of the subject of 'gene physiology,' at least, if not of gene physics and chemistry."[228]

At the same time that Muller's radiation experiments were in progress

131

another organism was being used for similar purposes. Lewis J. Stadler, using x-rays as well as radium, had exposed several series of germinating seeds of barley. "The barley plant produces several tillers from axillary buds, each tiller terminating in an inflorescence of about thirty self-fertilized flowers. In the dormant embryo the first three or four leaves are already differentiated, and the cells from which the tillers will be developed are separated. A mutation occurring in one of these cells, therefore, will affect only one tiller, and whether dominant or recessive, will segregate in the progeny of only one head."[291] There were no mutations among some 1300 tillers in the control; there were 3 mutations among 1000 tillers in the radium treated series; and 14 mutations appeared among 1200 tillers in the x-ray treated series. Stadler presented his results early in 1928, confirming Muller's discovery as well as demonstrating the effectiveness of the gamma radiation emitted by radium. Like Muller, Stadler hoped for an extension of radiation genetics to all of plant genetics, especially to maize, with its fairly well developed chromosome maps and series of multiple alleles.

At this time Muller was exploring numerous problems simultaneously; one of these was the cause of mutation itself. "If the change in a gene is not a direct effect, caused by the tremendous concentrated energy of a local electron 'hit,' but is produced indirectly through the intermediary agency of injurious chemical substances or physical conditions that become diffused through the cell as a result of the irradiation of the latter, then it becomes very likely that other chemical or physical treatments likewise will be able to produce mutations."[231] While the "indirect" effects could be suspected because of a slightly higher incidence of lethals in the untreated sperm of irradiated eggs, there was no evidence for the persistence of such an effect in later cell replications. No pigment changes in the individual eye facets were found, for example, whereas such minor sectoring should have been common had there been a persistent mutagenesis from the treatment of irradiated eggs. "This, by the way, strengthens the conclusion which I had previously reached that the tendency of a mutation to appear in only one half of a fly derived from treated sperm, instead of in the whole fly, indicates not that the mutation is delayed to the two-nucleus stage of cleavage but rather that the chromosome may be split already in the mature spermatozoan."[231]

Another problem was whether all mutations occurring spontaneously could be attributed to cosmic radiation, background radiation from the surface of the earth, and the intrinsic radioactivity of the isotopes of phosphorus and other trace minerals in the organism. This was ruled out, however, when a comparison was made by Muller and Mott-Smith of the radiation dose received by the flies from these natural sources and the frequency of spontaneous mutations. Spontaneous mutations were far more abundant than could be attributed to the sum of the radiation from all these sources. "The ordinary natural radiation will account for only about one one-thousandth of the natural mutations that occur in flies. . . . We must, therefore, conclude that practically all the mutations that occur in untreated individuals of Drosophila are produced by some other cause or causes than the natural radiation present in the general environment."[231] This suggested, once again, that some indirect, probably chemical, cause was involved, resulting in a minor intragenic change of the individual gene.

This nearly complete parallel of radiation effects to spontaneous mutation was disturbed, however, by several unexpected findings. High doses of radiation produced a large number of "dominant lethals," resulting in reduced yields of progeny; but spontaneous dominant lethals were much rarer in comparison to the spontaneous recessive lethals. Many of the F_1 progeny, apparently normal in phenotype, were sterile, particularly among the males. This was again higher in proportion to the induced recessive lethals than had previously been observed for the spontaneous mutations. And, as Muller and Altenburg had proved in 1930, the frequency of translocations was comparable to that of the lethals themselves at high doses. The ever-sporting displacements added another new phenomenon to radiation mutagenesis that was unknown for the spontaneously arising mutations. The lethality of many of the translocations when homozygous also suggested that x-rays caused mutations near the site of breakage or, at least, that their mode of action was more diffuse than that of the highly punctiform and localized effects of the spontaneous mutation process. The viability and apparently normal phenotype of many of the translocated progeny ruled out the possibility that the chromosome was a "continuum." "If the genetic material be regarded as a continuum, instead of as segmentally arranged in units, the genes, there would be no distinction between an intragenic and intergenic break, and all breaks would partake of the nature of gene-mutations. . . ."[247]

While the chromosome could still be considered segmented, the effects of radiation could no longer be considered identical to those involved in spontaneous processes. "It is, therefore, likely that there is a real difference between X-rays and the 'natural' cause or causes of mutation, in regard to their effectiveness in altering genes and in causing breakages and reattachments of parts of chromosomes."[247] The difference, however, was not so great that x-radiation would be set aside as a curiosity. There were other parallels of radiation-induced and spontaneous mutation: the recessive lethals were much more frequent than visible mutations; their localizations resulted in similar linkage maps; the induced mutations showed allelism to known spontaneous mutations; and the suggestive evidence of indirect radiation effects provided a chemical basis for some of the induced mutations. All these points convinced Muller that radiation exposure would duplicate the types of changes brought about spontaneously.

Stadler, however, had discovered quite another story. The consequences of radiation exposure were strikingly different from what was obtained spontaneously. So vast were the differences that he could not accept any similarity between the two processes, or at least the parallel between them was considerably less than appeared to be present in Drosophila. This was true for barley as well as maize. Stadler questioned the extension of Muller's interpretation of the Drosophila results to the induction of true gene mutations in plants. "The results of genetic experiments with x-rays in plants are not entirely in harmony with this view. Radiation induces gene mutation as well as grosser chromosomal variations in plants as in the fruit fly, and the induced mutations meet all tests of typical gene mutations. The evidence from plants considered alone, however, does not permit any sharp differentiation between the induced gene mutation and various extragenic alterations which may be expected to accompany the types of chromosomal derangement brought about by the treatment."[292] The x-ray mutations formed a special class of mutations, a "mechanical" class consisting chiefly of the deletions of one or more genes. This "mechanical" class, however, did not appear among the spontaneous mutations since the types of mutants which occurred there (especially the dominant mutations) were found rarely, if at all, among the mutations induced by radiation exposure. Muller's inference that gene mutations involved some sort of chemical change in part of the molecular composition of the gene was only conjectural. "Since chemical changes in the gene are beyond direct investigation, the conclusion that the genes are chemically transformed must be based almost entirely on negative evidence. To state that an induced variation is a gene mutation is not to explain it but merely to label it."[292]

The arguments Stadler brought forth included some novel suggestions. A few of the apparent gene mutations induced by radiation in pollen were viable as haploids in the gametophyte generation, but most of the gross chromosomal deletions and any gene mutations lethal to the gametophyte were screened out by the gametophyte. Were these survivors then true gene mutations? There were two reasons why Stadler rejected that possibility. First, McClintock had demonstrated the cytological existence of some deletions which were viable in the gametophyte. Second, "if these mutations are due to the loss of germinal material rather than to some change within the gene, the losses are so slight as to have no lethal effect on the haploid gametophyte."[242] Most telling, however, in barley, was the absence of the dominant types of mutation. "The most striking difference between the results in barley and the results of comparable experiments in Drosophila is the absence of induced dominant mutations in the plant. Why do x-rays induce dominant and recessive mutations indiscriminately in the fly, and induce recessive mutations only in barley?"[292] The relation of mutation to dose was linear for barley, and the relation of induced deficiencies to dose for pollen was also linear! Thus quantitative studies of the frequency of mutations could not be used to distinguish gene mutations from the class of events most likely to mimic them—deficiencies!

The combination of lethality, impaired fertility, and impaired viability which was characteristic of mutations induced by x-rays was in marked contrast to spontaneous mutation. "The mutations of these genes have no lethal effect on the gametophyte. Among more than one hundred mutant plants examined, none was affected by genetic partial sterility. Closely linked genes do not mutate together."[292] Hoping to use maize for a more detailed study of the events affecting the individual gene, Stadler, like Muller, saw the importance of irradiating known loci which have multiple allelic series, such as the aleurone and plant color series at the R and A loci. "Are the induced mutations of R^r, like the natural mutations, limited to one of the component genes or gene fractions, or do they, like deficiencies, usually affect both components together? These and many similar questions require evidence from the mutation of specific genes; they are untouched by experiments on the general rate of mutation."[292]

To Stadler's surprise six of seven loci selected for analysis, including the R locus, failed to produce any mutations after x-ray treatment. Only the waxy locus gave a few mutations (2 among 50,000 cells compared to none among 1.5 million for the spontaneous control). "The results of this necessarily small experiment lend no support to the assumption that mutation in general is 'speeded up' by irradiation. Induced mutation does not differ from natural mutation merely in its greatly increased frequency of occurrence; apparently there is a qualitative as well as a quantitative difference. . . . Most of the induced mutations in plants are due to various extragenic alterations, chiefly non-lethal deficiencies. This is true in spite of the fact that the variations recorded as mutations in the experiment with plants are exclusively the 'visible,' the 'lethals' being automatically excluded by the interposition of the haploid gametophyte generation."[292] Muller's argument that x-ray induced mutations can be induced to revert by radiation did not affect Stadler's conclusion. He proposed that a deficiency can revert if the fragment is still "floating around" in one of the cells. He did not believe that a fragment deleted from a chromosome is immediately digested or destroyed; it might eventually become reattached to one of the chromosomes in a subsequent mitosis whenever another break occurred. "The results," Stadler concluded, "do not support the assumption that mutation in general is affected by irradiation."[292]

In these early years of the 1930's, Muller was also finding that the case for gene mutations could not be built on any one single line of experimental evidence. The recessive lethals, as his student Oliver had demonstrated, were induced at a frequency directly proportional to the radiation dose administered. This was so regardless of the amount of time the treated sperm were stored in the inseminated females, although different stages of germ cell formation did show different sensitivities. For the translocations, the proportions were different with respect to dose from those for recessive lethals. It was thought that if the breaks were accompanied by the mutual exchange of chromosome arms of non-homologous chromosomes (the most typical form of translocation), then the two breaks required, if independent of each other, should give a translocation frequency that was approximately the square of the probability of each break being induced. Four times the

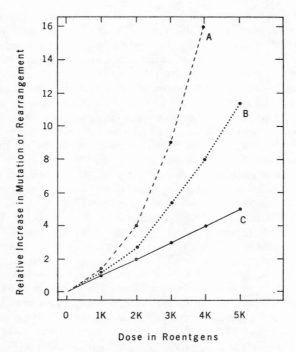

Figure 21. *The Linear and Exponential Radiation Laws.* In curve C, the proportionality between dose in thousands of roentgens (K) and mutation frequency is linear. This has been extended from a fraction of a roentgen to more than 10,000 roentgens. In curve A, the theoretical expectation for two independent breaks of a chromosome is proportional to the square of the dose. Since rearrangements involve two or more breaks, the actual curve obtained (B) falls short of this expectation. Differential viability of rearrangements and simultaneous multiple breakage have been proposed to account for the 3/2 power increase for this relationship.

dose of radiation should give sixteen times the number of translocations. But experimental results showed that the actual frequency was lower; about eight times the number were obtained by a fourfold increase in dose. This meant that the exponent was closer to 3/2 than to the square of the dose. It suggested, at first, that not all translocations were independent breakage events. To account for this, Muller suggested that the two non-homologous chromosomes might be in contact with each other at several points along their lengths in the sperm head and that a "hit" at such a contact point might induce a type of non-homologous "crossing over" of the chromosomes. In any case, the frequency of rearrangements was clearly different from that of the recessive lethals, and this was used as supporting evidence for the difference between point mutations and rearrangements.

In 1935 there were a number of new findings in *Drosophila* studies. Muller discovered a viable deficiency for the genes yellow and achaete.

The deficiency was genetically produced by a recombination between two chromosomes carrying similar inversions. In one of these inversions the left end tip of the chromosome contained the normal alleles for yellow body color and achaete bristle pattern with a break immediately to the right of achaete. The other inverted X chromosome was broken just to the left of yellow, so that the normal alleles of yellow and achaete were brought into contact with the right end of the X. A crossover between these two inversions resulted in one chromosome carrying both the left part (bearing the y^+ac^+) of one inversion with the right part of the other inversion (bearing the y^+ac^+). Consequently, the other product of crossing over carried no representative of the yellow and achaete loci. If such a chromosome lacking these two genes were fertilized by a Y-bearing sperm, it was expected that the male receiving it would be killed by the deficiency for these two genes. To his surprise, however, Muller found that these males were viable, having an extreme yellow and achaete phenotype! "The question must finally be considered as to why individuals deficient for two loci should be able to live at all."[235] If the entire theory of the significance of the gene as the basis for life and evolution was not to be discarded, then a much greater complexity had to be invoked for the relation of genes to characters. Thus there had to be a "prevalence of compensating systems and the overlapping of functions in general." Alternatively, "the particular genes in question are less basic and perhaps simpler in function than most genes—particularly could this be true in the case of a gene having to do with pigment production in an artificially protected organism."[235] Whatever the evolutionary significance might be, however, one important fact was established: deficiencies need not result exclusively in lethal effects, and visible non-lethal recessive phenotypes might exist which have demonstrable deletions of chromosomal material.

A second fact, established for Drosophila by Belgovsky, was the linearity of the frequency of minute structural rearrangements with the radiation dose. This made the apparent distinction between gene mutations and extragenic breakage phenomena no longer tenable, as Stadler's studies with maize had pointed out a few years earlier. Similarly, the analysis of several yellow, achaete, and scute mutations indicated a large number of structural changes, many of them very small; these were confirmed, cytologically, when salivary chromosome techniques were developed in the mid-1930's. The paradox that two breaks were required for both types of structural rearrangements, with gross rearrangements obeying an exponential law and minute or small rearrangements obeying a linear law, had to be resolved. If the radiation produced a cluster of ionizations along its path, then the probability of two nearby breaks being induced would no longer be exponential, since the breaks would not be independent, but would be a consequence of one primary x-ray. It was also possible that the degree of coiling and looping was more pronounced for small regions of the chromosomes than would be the case for gross contacts of non-homologous chromosomes. If so, single paths of breakage at the points of loop contact would be found more frequently among minute than gross rearrangements.

Despite these various restrictions on the experimental evidence for gene mutations in Drosophila, Muller rejected the possibility suggested by Goldschmidt that all mutations, of any origin, were different degrees of minute structural alteration. He also rejected Stadler's view that the "mechanical" process was the only mutagenic route possible for ionizing radiation. Goldschmidt's argument that all mutations were a consequence of multiple breakage and "pattern reestablishment" in a single continuous thread did not account for the lack of a relation between the effects of pattern rearrangement and the number and type of rearrangements. Thus a very small rearrangement might be lethal or produce pronounced visible defects, but many gross rearrangements, such as translocations and inversions (some with three or more breaks), would produce no severe abnormality. If a continuum did exist, what experimental value would such a model have if it could not interpret this independence of the genetic expression from the magnitude of the pattern change? The gene theory, on the other hand, did provide a predictive model for this paradox. The position effect on certain genes indicated a functional relation between genes. The genes themselves, however, retained most of their functional and structural discontinuity in the chromosome. The elimination of background radiation as the cause of most spontaneous mutations in Drosophila made it implausible that the same ionizing mechanism of breakage would be operating naturally, especially if the causes were chemical. Such evolutionary considerations in favor of chemical changes of the gene to provide for its progressive differentiation, were still indirect. Five years of intense radiation analysis of Drosophila failed to yield the positive demonstration of gene mutations.

The lack of direct proof of gene mutation in Drosophila and the absence of any experimental evidence for point mutations in irradiated plants made it improbable that Stadler's views would be rejected. Using the operational procedure of screening for gross and minute rearrangements, Stadler classified the residue as point or gene mutations. But "mutations are thus by definition alterations of unknown nature."[293] This did not prevent Stadler from acknowledging that they exist, but he still maintained that they were not produced by x-rays. "The results with maize (in which there are clear differences between x-ray and spontaneous mutations) do not indicate that this is true of mutation in general, but rather that the x-ray mutations are a special class. In Drosophila differences between x-ray and spontaneous mutations have not been obvious, and Goldschmidt, . . . assuming that the x-ray mutations are representative of the phenomenon of mutation in general, is inclined to dispense with the notion of intra-genic variation altogether."[293]

The differences between maize and Drosophila were significant. "We have at present no evidence of the occurrence in maize either of position effects or of deficiencies or mutations at translocation points."[293] To illustrate the mechanical theory of radiation damage in maize, Stadler compared mutations induced at the A locus by x-rays with those induced by ultraviolet irradiation (which is non-ionizing). The results from the x-ray treatments were discouraging for Stadler. "Loss of the A effect may be identified in the seedling stage, and it is therefore possible to make the determination on large numbers. We have identified about two hundred of such seedlings. Those due to deficiency of A or inviable mutations should show corresponding defects in fertility and development; those due to wholly viable mutation should be normal plants of full fertility, except for the coincident occurrence of derangements induced at other loci. The errors from coincidence may be avoided to a considerable extent by the use of low doses. . . . Among the A losses so identified, the great majority were found to be distinctly defective plants which failed to reach the flowering stage. Among those which survived to flowering, the majority were visibly defective in development and in all but a few the defective pollen for which they were segregating was of the extreme aborted type. . . . Those not visibly defective in plant development included cases with pollen of varying degrees of defectiveness. The last class, plants with no visible change except loss of the A effect, may be regarded tentatively as mutant a genes resulting from x-ray treatment. They constituted about two percent of the 'A losses' observed. . . . The 'mutant a' in . . . two plants of this type . . . was recovered and crossed with A. The heterozygous plants thus produced showed in both cases full transmission through female germ cells but much reduced transmission through male germ cells. Thus this extensive trial failed to yield a single case of mutation to the typical recessive a."[293]

The ultraviolet irradiation of pollen gave a different pattern of mutation. "Three cases of mutation apparently to the recessive a phenotype and one case of mutation to a new intermediate phenotype designated 'light,' A^{lt}," were obtained. "These mutants are wholly normal in growth and in pollen development. All four have been tested in male transmission, and all are normal."[293] Stadler's interpretation of the results, like Muller's could be anticipated: "Insofar as they are comparable, they suggest a basic difference in the nature of the mutations induced by the two agents."[293]

Stadler's Paradox

Shortly before his death in 1954, Stadler prepared a critical evaluation of the gene concept. Its posthumous publication was considered his valedictory address on the gene. At the time of its writing the profound discoveries of molecular genetics had not yet begun. The gene concept was still derived from the experimental and theoretical studies of Drosophila and maize. Coming, as it did, forty-five years after the word "gene" was introduced in an undefined state, Stadler's criticism presents the most re-fined distillate of the attacks on the gene. The article does not attempt to repudiate the usefulness of the concept; rather, like Morgan in 1910, Stadler attempted his analysis "in the spirit of the judge, rather than that of the advocate."

To the student of molecular genetics, most of Stadler's criticisms, by hindsight, are invalid. But it illustrates why scientific controversy may wait twenty or more years for its solution. Stadler, using the remarkable property of genetic material to reproduce its variations, singled out the properties of the genetic substance from all other known physical phenomena. "The genic substance . . . appears to have properties quite different from those with which we are familiar from our knowledge of the physical science of non-living matter. Modern physical science gives us no model to explain the reduplication of the gene-string in each cell generation or to explain the production of effective quantities of specific enzymes or other agents by specific genes. The precise pairing and interchange of segments by homologous gene-strings at meiosis also suggest novel physical properties of this form of matter. These facts indicate that a knowledge of the nature and properties of the genic substance might give clues to the distinctive physical mechanisms of life. . . . The difficulties in the study of the genic substance are obvious. It cannot be isolated for chemical analysis or pure culture. The possibility of direct analysis of specific segments or individual genes is, of course, even more remote. The properties of the genes may be inferred only from the results of their action."[294]

This pessimistic view, partially justified at that time, forced Stadler's attention to the types of activity of genes themselves. Foremost among these was that of mutation. But caution had to be used not to confuse experimental observation of induced gene mutations with the model of mutagenesis favored by the experimenter. "When we conclude from an experiment that new genes have been evolved by the action of x-rays, we are not simply stating the results of the experiment. We are, in the single experiment, combining two distinct steps (i) stating the observed results of the experiment, and (ii) interpreting the mutations as due to a specific mechanism. It is essential that these two steps be kept separate, because the first step represents a permanent addition to the known body of fact, whereas the second step represents only an inference that may later be modified or contradicted by additional facts. When the two steps are un-consciously combined, we risk confusing what we know with what we only think we know."[294]

Some forty years earlier the Drosophila group had advanced strong reasons for separating the transmissible unit from the observed character in the term unit-character. Its confusion had led Castle astray for some fifteen years. Now Stadler was placing gene mutation in the same category. "The mischief involved in the use of the same term for the two concepts is obvious."[294] The term gene mutation, however, was dependent on the gene concept that a particular investigator had in mind. "The significant ambiguity is not in our definition of gene mutation but in our definition of the gene itself, because any definition of gene mutation presupposes a definition of the gene."[294]

The definition which Stadler offered was based on Bridgman's philosophy of "operationalism." "The essential feature of the operational viewpoint is that an object or phenomenon under experimental investigation cannot usefully be defined in terms of assumed properties beyond experimental determination but rather must be defined in terms of the actual operations that may be applied in dealing with it."[294] Using Bridgman's operational viewpoint, Stadler offered a definition of the gene. "Operationally, the gene can be defined only as the smallest segment of the gene-string that can be shown to be consistently associated with the occurrence of a specific genetic effect. It cannot be defined as a single molecule, because we have no experimental operations that can be applied in actual cases to determine whether or not a given gene is a single molecule. It cannot be defined as an indivisible unit, because, although our definition provides that we will recognize as separate genes any determiners actually separated by crossing over or translocation, there is no experimental operation that can prove that further separation is impossible. For similar reasons, it cannot be defined as the unit of reproduction or the unit of action of the gene-string, nor can it be shown to be delimited from neighboring genes by definite boundaries."[294] Using this operational approach, Stadler hoped to augment the concept of the gene, because the term itself contained the ambiguity that Johannsen hoped could be avoided by its introduction. "The term *gene* as used in current genetic literature means sometimes the operational gene, and sometimes the hypothetical gene, and sometimes, it must be confessed, a curious conglomeration of the two."[294]

But if the term *gene* could be defined operationally through a "specific genetic effect," could this same operational procedure be used to define gene mutation? Stadler said no. "To say that no operational definition is now possible is only to repeat in different words . . . that we have no positive criterion to identify mutations caused by a change within the gene."[294] Stadler's pessimism was based on the "general" class of mutations, such as lethals or visibles, that was frequently used for exploring problems of mutagenesis with radiation and other agents. To try to sift through all these events in the chromosomes was hopelessly confusing and time-consuming. Stadler therefore urged a different approach—a return to the study of individual genes. "The chief advantage in focusing study on the single gene is that this makes it possible to substitute the direct experimental analysis of specific mutants for the application of generalizations assumed to apply to mutations at all loci."[294]

In applying this to his own studies of the R and A loci in maize, Stadler found that it was not possible to rule out a "compound gene" because position effect and multiple allelism could be explained with such a model as readily as without it. "The notion of the compound gene, or some equivalent unit, may prove to have significance, since there may be special relationships among the clustered elements that mark them off as a group from adjoining unrelated elements. One of these may be the interrelationships in gene action between the clustered elements, which could lead to the occurrence of position effects when members of the cluster are separated by crossing over or translocation. This may be a basic factor in the explanation of position effect in general."[294] Similarly, "although different alleles may have widely different numbers of genic elements, none is actually a deficiency. In terms of the postulated origin of the cluster, all of those with more than a single element may be considered duplications. On the other hand, when we arbitrarily take as the standard type an allele carrying several genic elements, other alleles with fewer elements will appear as deficiencies, and the mechanisms that produce them as mutants from the standard type will be mechanisms of gene loss."[294] As in the Bar case, before the cytological discovery of its structure, some geneticists had considered the reversion of Bar to normal to be merely the loss of a new "presence." This hypothesis was abandoned when Bar was shown to be a duplication. Subsequent analysis of such reversions showed them to be either a loss of one of the duplicated segments or an alteration in one or both of the segments, leading to a destruction of the position effect existing between these segments. Stadler's criticism extends the same principle to series of multiple alleles affecting a character quantitatively. Mere change in character does not imply a corresponding loss or gain of genetic material.

These cogent arguments, based on the operational viewpoint applied to the gene in maize genetics, presented a paradox: "In the study of gene mutation, we are for the present in an anomalous position. A mutant may meet every test of gene mutation, and yet, if it is not capable of reverse mutation, there is ground for the suspicion that it may be due to gene loss, while, if it is capable of reverse mutation, there is ground for the suspicion that it may be due to an expression-effect. The only escape from this dilemma is through the more intensive study of the mutations of specific genes selected as best suited to detailed genetic analysis, in the hope of developing more sensitive criterions for the identification of gene mutations."[294]

Stadler's paradox, ruling out reverse mutation as decisive for describing gene mutation, left the gene concept where it began: undefined. While this may have been true at the time for maize, the techniques for further exploration of this problem in *Drosophila* were not exhausted, and completely unexpected approaches to the gene concept, in other organisms, were already being proposed when this stimulating document was posthumously published.

CHAPTER 17

Discontinuity: "Step-allelism" Versus the "Left-Right" Test

While Morgan, and the Drosophila group in general, rejected the presence and absence theory on the grounds of its oversimplified interpretation of allelism, Bateson, to the time of his death in 1926, did not agree that the theory was in any way weakened. For one thing, he never contended that the absence of a character was identical to the loss of the entire transmissible element for a multiple allelic series. "The representation has the advantage of extreme simplicity. It implies only that something, whether a material something or otherwise, is present in the one . . . which is absent from the other. . . . With the supervention of the chromosome theory, for reasons which I have never clearly understood, the interpretation based on Presence and Absence was set aside. . . . The argument most frequently specified as militating against Presence and Absence is derived from the phenomenon of multiple allelomorphism. . . . Rightly regarded, however, I contend that the multiple series not merely contribute nothing capable of such construction, but that their existence amounts almost to a demonstration of the correctness of the Presence and Absence hypothesis."[23] Bateson's hypothesis was based on the parallel of the quantitative coloration in the white series and an inferred quantitative loss of the components of the normal allele for the series. "Factors composing a multiple allelomorphic series produce an effect in degrees quantitatively different. This is

143

exactly what is expected. The factor put in on the one side is absent, wholly or in part, from the other, and, on segregation, the two qualities which respectively were combined in fertilization, reappear. Why Morgan should declare that 'only one kind of absence is thinkable,' I do not understand. We should not assert that because a sovereign is absent from a purse, that purse must contain nothing."[23]

A year after Bateson had died, A. S. Serebrovsky, stimulated by a visit to the U.S.S.R. by Muller in the early 1920's, began studies with *Drosophila*. He noticed a difference in the map lengths between certain genes when they were in the coupled arrangement rather than in the repulsed arrangement. This was particularly true if three or more genes were simultaneously used in a linkage analysis. His interpretation, unlike Castle's, did not reject the theory of linear linkage; instead he maintained that theory, but assumed that the variability of the map size was caused by the asymmetry that would result if the mutations involved were deletions of varying size. With all the deletions coupled in one chromosome, the possibility for crossing over would be more restricted and would give a smaller map size than if distributed equally among the chromosomes used. "It is not difficult to see that the results obtained . . . can be used for the resurrection of Bateson and Punnett's Presence and Absence theory, which seemed buried. . . . Our opinion is that the Presence and Absence theory can be brought into agreement with the Chromosome Theory of Heredity. Only both theories must give up the postulate of the indivisibility of the gene. If the results obtained in the present investigations are confirmed by investigations on other genes, it will perhaps be possible to explain the series of allelomorphs by the hypothesis that in different mutations parts of chromosomes unequal in . . . size disappear."[274] Thus, if white represented a loss of 0.10 map units, ivory would represent a loss of 0.05 map units and eosin a loss of 0.03 map units. An extension of this hypothesis to several sex-linked genes at the left end of the X chromosome made this theory seem even more plausible to Serebrovsky.* But to obtain more convincing evidence, Serebrovsky and his students began to irradiate flies, using Muller's new techniques, and they found several changes at the very tip of the X chromosome which affected bristle development. One of his assistants, N. P. Dubinin, in crossing these to one another, noticed a surprising allelic relationship.

In the normal fruit fly there are a number of larger bristles located on the "back" or thorax of the fly, as well as somewhat smaller ones located on the head and on the ventral surface of the thorax. In addition there are extremely small bristles, microchaetae, which form a series of parallel rows along the length of the thorax. Various mutations at the tip of the X

---

* Serebrovsky's results were probably a consequence of the different modifiers in the stocks he used. This difficulty was avoided by Muller and Raffel in their left-right analysis on several similar scute mutations (see p. 153). They made sure that the mutant alleles were in comparable genetic backgrounds.

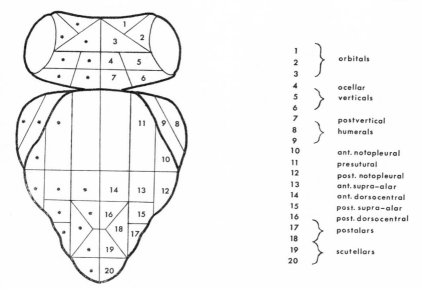

1 ⎫
2 ⎬ orbitals
3 ⎭

4   ocellar
5 ⎬ verticals
6

7   postvertical
8 ⎬ humerals
9

10   ant. notopleural
11   presutural
12   post. notopleural
13   ant. supra-alar
14   ant. dorsocentral
15   post. supra-alar
16   post. dorsocentral
17 ⎫
18 ⎬ postalars
19 ⎬ scutellars
20 ⎭

Figure 22. The Location of the Scute Bristles. The numerous bristles were believed to be controlled individually (or in groups) by subgenes of the scute region. Some thirteen such subgenes were ordered by complementation mapping.

chromosome affect different numbers of these bristles or of the microchaetae. These bristle mutations were first noticed on the shield-shaped scutellum of the thorax and consequently the series was designated as the *scute* bristle series. Close to the bristle forming region was another gene whose mutant alleles expressed yellow body color. Some of the yellow mutations showed bristle or microchaete abnormalities (achaete). Using the first few scute and yellow mutations induced by radiation, Dubinin was able to demonstrate that yellow generally does not extend its influence to scute in the compound y/sc; but various scutes in compound showed only those bristles which each of the scutes had in common. This was not a new phenomenon among alleles; Muller had recognized, ten years before, that in the truncate (dumpy) series, the mutants oblique wings (o) and thoracic vortices (v) were wild-type (+) in compound although each of these mutants was viable and manifested its character when in compound with truncate (olv) or dumpy (ov). The modern geneticist would designate this behavior as *complementation*, but there was no single term used to express this allelic phenomenon at that time. "Cases now known are fairly numerous in which different recessive mutant allelomorphs of the same locus have effects which are to some extent or almost wholly, different in their character or in their location on the organism. Thus, mutant allelomorph 1 may affect character A very much and B very little or not at all, while

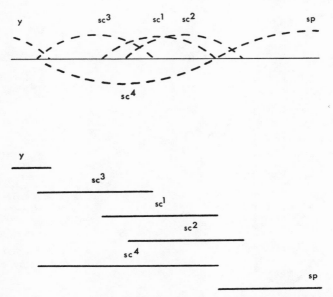

Figure 23.   An Early Complementation Map of the Scute Region. The four scute mutations, yellow body color, and a spoon-shaped wing mutation were ordered (top map) by complementation mapping in the order shown. The contemporary procedure for representing a complementation map of these alleles is shown in the lower map.

allelomorph 2 affects A little and B much. Such allelomorphs, when crossed, usually form a compound that is more normal than either. For, in respect to each character effect or body region, the more normal effect is usually the more dominant; that is, the compound is usually in respect more like that allelomorph which has a more nearly normal effect on that character or region."[233]

The bristles which were expressed in the compounds suggested to Dubinin and Serebrovsky a model that extended and eventually replaced the presence and absence theory in their interpretation of gene structure. If the gene were composed of a certain number of segments or subgenes, each of which controlled the formation of one or more of the bristles on the fly, then each new mutation would very likely affect different subgenes within the "basigene" for scute. Any two alleles in compound, if lacking mutual subgenes, most probably would demonstrate a region of overlap for the topographical defects shared in common by the two alleles. If a third allele were tested with each of the other two, then a further set of overlapping regions would be demonstrated and the various overlapping patterns could be arranged in a series of continuous steps, extending from the particular scute allele with the leftmost subgenes lacking or altered to that with the rightmost defective subgenes. The individual alleles, then, could be called step-alleles. The process of step-allele formation was called transgenation, and the theory itself was called the "step-allele" or "subgene" hypothesis.

The discovery of a fourth scute allele, however, presented difficulties to this linear plan. I. J. Agol did not believe this detracted from the utility of the step-allele model. "We realized that it was impossible to place all the known segments in a linear series. Before the discovery of the scute[4] transgene all the segments of the step-alleles of the scute type were thought to be arranged in one line. . . . However, due to the . . . study of scute[4], a number of facts have come to light which did not agree with this assumption. Not all the segments of the step-alleles which we studied can be arranged in a line. . . . We came to the conclusion that the segment affecting the anterior dorsocentral bristles is located to the side of the yellow-scute[4] line. . . . The scute basigene which we studied does not have a linear but at least a two-dimensional structure. One assumes that further studies in this direction will lead to a three-dimensional structure of the gene."[1] The fact that some scutes were associated with yellow, others with a lethal, and still others with a defective spoonlike curving of the wings, made it seem unlikely that clear-cut boundaries existed between one gene and another. At least the alleles, on this model, did not appear to stay contained within the confines of single genes. "All the other alleles, as well as the so-called repeated mutations, should be carefully studied. Theoretically we may assume a long series of step-alleles passing gradually from one gene to the other along the entire length of the chromosome. Two different groups, not allelic to each other, may be partial alleles of one another and the same gene. . . ."[1] Such an overlapping of one basigene into another would suggest that the chromosome as a whole may be a continuum. "If the whole chromosome is a continuum, i.e., if one gene passes directly into another, then crossing over may take place only at the gene and not between genes, since step-allelism excludes any intervals between genes."[1]

About two years later there were some thirteen different scute alleles which affected this region; their phenotypes included body color, viability, bristle and microchaete patterns, and wing shape. Dubinin tried out the various compounds between different pairs of these scute alleles and restored the map of the basigene to a linear array. The "topography of the . . . scute basigene" contained some thirteen "centers" (or "complementation units" as they would be called today). They were defined by the thirteen different "groups of centers" (today designated as "complementation patterns" or "line segments") expressed by these alleles.

The model had predicted that if two alleles were found affecting "centers" that did not overlap (i.e., form a step) the fly would be normal in phenotype. This was borne out: "Scute[3] alone presents a link between the two independent groups of scute and achaete transgenes. . . . If we had not had scute[3] at our disposal, we should have been obliged to recognize these two groups of genes as two independent, differently located series of multiple allelomorphs though inseparable one from the other by the crossing-over method. . . ."[117]

The step-allele theory was criticized by several investigators using widely different approaches. Sturtevant and Schultz, in 1931, used modifiers of bristle number to test the map relations of the various scute genes. The effects of the modifier, Hairless, were contradictory to the expectations on the step-allele map. "Achaete lies to the extreme left of Serebrovsky's map, scute-1 covers most of the right half. The scute-1–achaete heterozygote is completely normal; but both act as allelomorphs to scute-3, which on Serebrovsky's interpretation is to be taken as resulting from mutation of practically the entire series of subgenes. We have studied scute-1 and achaete in flies that were also Hairless (the gene concerned is a III-chromosome dominant that removes bristles in a pattern different from that of the scutes). . . . The data given leave no alternative to the conclusion that both scute-1 and achaete are acting on bristles scattered throughout the entire length of Serebrovsky's map, though many of these effects, under ordinary conditions, are of subthreshold degree. . . . This conclusion is in harmony with most recent ideas concerning the way in which genes control development. It would seem to us to destroy the value, as a working hypothesis, of the extreme preformationist subgene hypothesis. If each scute subgene is affecting many or all of the bristles of the fly, the whole logical basis of the hypothesis is undermined."[206]

A few years later Child demonstrated that temperature would affect the patterns obtained for the different scute alleles, indicating, as did Sturtevant and Schultz, that the scute region was a unitary rather than subgenic region in its physiological response.

While these controversies were being explored at a genetic level, two new cytological techniques were coming into usage. J. Belling, in a series of papers between 1928 and 1933, proposed a model of gene structure in plants that he hoped would be consistent with cytological phenomena such as crossing over, chromosomal rearrangement, and mutation. In the genera *Lilium* and *Aloë*, Belling developed a technique which revealed extremely detailed chromosome structure during the prophase of the meiotic reduction division. The smallest elements he could observe were *chromioles*. When the homologous chromosomes were closely paired in the pachytene stage, these chromioles showed a discontinuous clustering. Each cluster he designated as a *chromomere*. "It is proposed to designate as chromioles the two spherules of each side, which appear more closely connected than are the two pairs. Thus two pairs of chromioles make a bivalent chromomere."[35] Using one segment of a pair of chromosomes ("bivalent"), Belling attempted to calculate the total number of chromomeres in the nucleus of a cell. "The number of ultimate chromomeres (sets of two pairs of spherules each) was calculated from the number in a measured distance of a camera lucida drawing where they were readily countable. Thus there were 65 ultimate chromomeres in 43.5 microns. Hence the total number of ultimate chromomeres in the cell was 2193."[35] This number was strikingly similar to Muller's calculations of the number of genes in the fruit-fly genome. Also, the homology of the chromomeres, their varying sizes, and their fixed linear order in any specific chromosome, were attributes characteristic of hypothetical genes. "Correct scientific procedure, in the writer's opinion, demands the adoption, as a working hypothesis until a better one appears, of the assumption that chromioles and chromomeres are genes, doubtless with more or less of an envelope."[35]

When Painter introduced salivary gland chromosomes as material suitable for cytogenetic studies in the fruit fly, Koltzoff and, independently, Bridges, offered an interpretation of the unique structure of banding found along the length of these chromosomes. "According to the theory of Koltzoff—which we accept, the nodes or rings of the salivary gland chromosomes represent chromomeres, inasmuch as an entire chromosome in a salivary gland cell is a hollow cylindrical bundle of uncoiled and parallel-lying chromonemata, formed by the repeated division and conjugation of the single chromonemata in two original homologous chromosomes, and since homologous chromomeres of all the chromonemata of a bundle would lie opposed, they would give the appearance of rings or cross-striations. . . ."[249] The salivary chromosomes showed bands which were well within the visible limits of optical cytology; thus if they corresponded to genes or groups of genes, the calculation of gene size and number could be reinterpreted from an entirely new approach. Similarly, if genetic analysis of the mutations affecting single genic regions were carried out, the number of extragenic and, to some extent, intragenic, rearrangements could be estimated. Most significantly, however, the cytological analysis of conclusions based on genetic tests could prove a powerful tool for confirmation or rejection of the conclusions. All these implications were tried out by Muller in association with A. A. Prokofeyeva. Of these, one technique stood out as the most critical test of the step-allele theory. It also provided a new basis for defending gene mutation rather than deletion or position effect as the chief mechanism of "natural" mutation.

The "Left-Right" Test

The number of problems that needed exploration in the mid-1930's seemed to accumulate faster than the contributions which were made by radiation analysis of the genetic material. The first task seemed to be that of exploring the structure of hereditary material itself. "Granting that the genetic material is not set together in a mechanical fashion, but forms an organic system in which the interrelation of the 'units' is as important in its way as the inner composition of the latter (these two features being in fact mutually dependent), it is the task of genetics to attack in a concrete way the problems of the actual nature of this system, the manner in which it is built up of parts, and these of sub-parts, their visible cytology in relation to their functional genetic expression, how and to what degree these parts may be separated and the effects of such separation, and conversely, the effects of the connections that normally exist between them."[248]

Muller's new approach was applied to a restricted region of the chromosome—the leftmost tip of the X chromosome—in which the normal allele for scute resided. Muller had found that most of the scutes used by Serebrovsky, Dubinin, and their colleagues were rearrangements of a gross nature, either translocations or inversions. The rearrangements were selected and classified, and "as many cases of breakage as possible were analyzed with reference to one another, in order to determine what limitations there might be in regard to the positions and manner of breakage."[248] The technique concentrated on the inversions in the X chromosome showing scute abnormalities. If two inversions, A and B, had breaks in the scute region, and if the breaks on the right end of the chromosome were close to each other, a direct genetic test could prove whether the breaks in the scute region were in the same place or not. This test, the "left-right" test, would produce two products of crossing over. One of these would contain the left-hand portion of A and the right-hand portion of B. The other product would be a crossover combining the left-hand portion of B with the right-hand portion of A. If the break in A were to the left of scute and the break in B were to the right of scute, then the inversion A would contain scute on the right side and inversion B would still retain scute on the left side. The recombinant with the left side of A and the right side of B would consequently lack scute altogether, and the recombinant containing the left side of B and the right side of A would have a duplication for scute, each scute residing on opposite ends of the recombinant chromosome. If the scute deficiency is viable, it should show an extreme scute phenotype, while the duplication should show a normal, or nearly normal phenotype.

Using this "left-right" test, Muller reported that "thus far seven breaks in the region in question have been analyzed. . . . Among these seven cases, only four perceptibly separate positions have been found."[248] Thus in several different inversion pairs, there was no difference between the two products of crossing over, as if the two breaks in the scute region were at an identical point. "The conclusion thereby becomes probable that there are only certain definite points at which the chromatin may ordinarily be broken by irradiation, and that, by the methods used here, it is possible to discover the totality of these points within a region circumscribed in the way explained. The blocks of material between these points may be regarded as 'genes,' and, if this be true, the distances apart of the breaks studied . . . are of the same order of magnitude as the individual genes, and thus the absolute number of genes contained within the restricted region is determinable. . . . At the same time no evidence has appeared that a gene—even the mooted gene for scute—can be divided within the limits of its own struc-

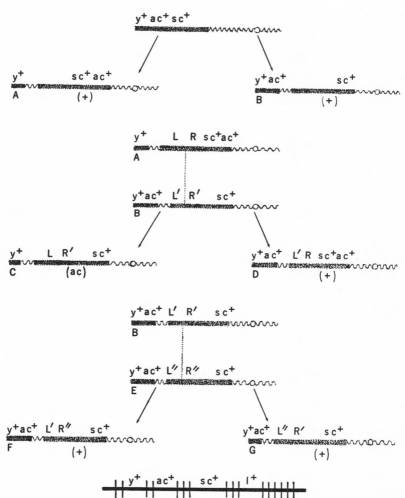

Figure 24. Muller's Left-Right Test. The normal chromosome bearing y⁺ac⁺sc⁺ differs in gene sequence from two rearrangements, A and B. In A, the left break has separated y⁺ from ac⁺ and sc⁺. The right breaks are in the heterochromatin (wavy line). Two rearrangements, B and E, have left breaks separating y⁺ac⁺ from sc⁺ and right breaks in the heterochromatin. Crossing over between A and B yields a deletion of ac⁺ in C and a duplication of ac⁺ in D. Crossovers between B and E yield no duplication or deletion in F and G for the genetic material in the y⁺ac⁺sc⁺ region. These left-right tests were used by Muller and his associates on sixteen rearrangements. Their break points indicated numerous points at which breakage could occur without loss or gain of material in a left-right test. These breakage points led Muller to believe that the chromosome was linearly discontinuous.

ture, in such a way as to permit the continued life of both separated portions."[248]

The actual dimensions of the gene as defined by this breakage test, could be measured from the cytological study of minute structural changes in this region of the X chromosome. The best of these extragenic alterations was the scute-19 case. The scute-19 deletion occurred when a small fragment was excised by radiation from the tip of the X; this same fragment was inserted near the dumpy locus of the second chromosome. This extremely small "transposition" was viable in homozygous condition. Also both males and females containing the duplication in the second chromosome were viable, with the scute-19 material functioning normally for the genes yellow, achaete, and a lethal gene to the right of scute. A position effect on the gene for scute, however, gave a reduction in bristle number. The deleted portion was lethal, but any chromosome fragment containing the normal allele of the lethal to the right of scute rendered it viable. It was thus small enough to contain not much more than four genes on the discontinuous model of the chromosome structure. "Prokofeyeva . . . found that the whole region occupies only a portion, about three-quarters, of one chromatic ring or node (the second) as seen in this material. Estimating the proportion which the region . . . forms of the total material present in all the chromatic regions, we can thus arrive at a tentative approximation of the number of genes in the chromatin; it is of the order of a few thousand."[248]

The conclusions from the work, then, restored the gene to its hypothetical status in 1927. "The above work indicates the potentially discontinuous character of the hereditary material, in that it is divisible into definite blocks capable of self-propagation in any new arrangement, and it also indicates that this divisibility has definite limits of size. On the other hand, studies of the phenotypic effects produced in the presence of these rearrangements have brought us strong evidence that the genes are, from the point of view of their functions in determining the characters of the organism, not discontinuous, in that neighboring genes enter into special relations with one another, with the resultant formation of a gene-system of a specific pattern, the joint functioning of which, as such, is necessary for the normal development and maintenance of the somatic characters. . . . The basis of genetic determination of the characters of an organism cannot be stated completely by a mere listing of the individual genes which the organism contains; the arrangement is itself, in effect, genic in nature, and the genetic material is in this sense a continuum."[248] Additionally, the cytological analysis of the y-ac-sc region, through the scute-19 segment in particular, made it possible to place a restriction on Belling's hypothesis. "The chromomeres contain in some cases at least whole groups or clusters of genes, the individual members of which are linearly arranged and are separable from one another both by breakage and by crossing over. . . . This conclusion seems at first contrary to the conclusion reached by Belling regarding the one : one correspondence between genes and chromomeres in

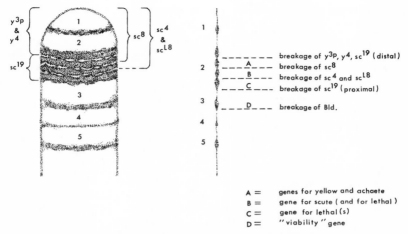

$$A = \text{genes for yellow and achaete}$$
$$B = \text{gene for scute (and for lethal)}$$
$$C = \text{gene for lethal (s)}$$
$$D = \text{"viability" gene}$$

Figure 25. Muller's Breakage Map of the $y^+ac^+sc^+$ Region. The four bands below (2) are broken in different places by the yellow and scute mutations shown in brackets. A single chromonema showing these break points defines the genes, A, B, C, and D, whose duplication or deletion results in a recognizable phenotype in the left-right test.

certain flowering plants, but it is possible that the genes are more regularly spaced and isolated in the latter. . . ."[249]

It is interesting that the calculation of the size and shape of the gene based on the breakage boundaries in the left-right test and the scute-19 salivary chromosome analysis are of nearly the same magnitude as those made earlier, before radiation analysis, on the basis of the Poisson distribution of alleles among the genes of the X chromosome. Comparing the uncoiled chromonema of a salivary chromosome to the condensed mitotic metaphase chromosome corresponding to it, Muller estimated that for the size of the gene, the "length is six to thirty times as great as its diameter. . . . This is in agreement with the fact that proteins and other complex organic molecules in general are chain-like, being much longer in one dimension than in the other two."[237]

In 1940 Muller and his associate, Daniel Raffel, carried out an elaborate study of three scute alleles whose left break points were at apparently the same place and whose right break points were in different parts of the heterochromatic segment of the X chromosome. This genetic similarity, inferred from the near identity of the phenotypes of the scute defects, was proved to be intrinsic to the breakage near scute. In Muller and Raffel's experiment, all three of the inversions were made co-isogenic with one another for their autosomes and for most of the sex chromosome between the break points. In contrast to these three cases, some twenty other scute alleles showed very marked differences. "The numerous rearrangements, although all appearing to possess one point of breakage and interchange of gene connections in one of four definite positions near scute, after the breakage and reattachments had taken place, were in all cases very different from those in the normal chromosome and . . . from each other. Correspondingly there was very much diversity in the expression of the 'alleles,' although no fixed relation was discernible between the map position of the different chromosome breaks and the kinds of scute phenotype associated with them."[267] The three similar scutes, sc[81], sc[L8], and sc[4] all had a cytological and genetic break point immediately to the right of the scute. Their other breaks were in the heterochromatic area; sc[4] had a break to the left of the bristle mutant, bobbed; sc[L8] had a break immediately to the right of bobbed; and sc[81] had its break somewhat further to the right of bobbed, in a region designated "block A." Using the left-right test on these three scutes in all possible combinations, Muller and Raffel confirmed the specificity of the scute effect. "These three mutations are, even when subjected to this more exact comparison, very similar to one another but . . . they are not identical."[267]

Several significant conclusions could be drawn from the experiments. "A proper evaluation of the meaning of our results now involves us in considerations of the degree of divisibility of the genetic material, the criteria by which we recognize the presence of genes and their spatial or numerical limits, and the extent to which the string of genes may be regarded as discontinuous or continuous in structure and functioning."[267]

Without designating his criteria as operational, Muller evaluated the various techniques for studying genes some fifteen years before Stadler emphasized the importance of exploring the gene at different operational levels. "In genetic theory, genes have been considered as (1) crossover units—hypothetical segments within which crossing over does not occur; (2) breakage units—again hypothetical segments within which chromosome breakage and reattachment do not occur (at any rate, not without destruction of one or both fragments); (3) mutational and functional units —those minute regions of the chromosomes, changes within one part of which may be so connected with changes in the functioning of the rest of that region as to give rise to the phenomenon of (multiple) allelism; or (4) reproductive units—the smallest blocks into which, theoretically, the gene string could be divided without loss of the power of self-reproduction of any part."[267] Muller's fourth criterion is a virtual paraphrase of the operational definition of the gene used almost fifteen years later by Stadler. Muller, however, did not attribute more than a theoretical or hypothetical existence to the assumed segments. "Although it seems often to have been assumed, there is as yet no empiric evidence, and only doubtful theoretical ground, for assuming that the lines of demarcation between genes, as defined on any one of these systems, would coincide with those of any of the others, or even for assuming, in the case of some given one of these systems (especially the mutational one) that such lines of demarcation are necessarily invariable, non-overlapping, well defined and absolute."[267]

The three scutes investigated by Muller and Raffel, and the additional analysis of ten other scute alleles, provided the same four points of breakage which were detected in 1934. Muller inferred from the total analysis that the breakage boundaries of this region were reproducible. "The most detailed studies have shown that not more than two entire bands, as seen with visible light, can lie between the leftmost and the rightmost of our breakage points (that is between the left and right breakage points of scute-19), and the ultraviolet photographs of Ellengorn . . . have shown not more than three bands. If we regard yellow and achaete as separate loci, and then take into consideration the evidence showing four different breakage points, we must conclude that there are at least four genes in this tiny region, hence more genes than detectable bands."[267] Also, "since we must thus admit that the original 'scute gene,' as defined by the mutational test, is really divisible into two parts, 'achaete' and 'scute' proper, as defined both by the tests of breakage and of crossing over, we should exercise caution before denying that the remaining 'scute' proper may be further divisible, even though the parts of it might not have nearly as completely differentiated functions as the earlier postulated 'sub-genes' were thought to have."[267] However, if subgenes of some sort were present inside the scute gene 'proper,' then the left-right test should have detected that such subgenes were missing between the two breaks near scute. "If these smaller deficiencies involve genes or 'subgenes' that form part of the hypothesized scute complex, and have their morphological expression mainly in bristle pattern, it might be thought that it should have been possible to detect them by reason of the relative absence of bristles in one of the classes of recombinants (the deficient class) as compared with the other class. Yet, as we have seen, there was no evidence of such complementary action."[267]

The results of the experiments could be applied to two problems: position effect and the structure of the gene. The similar scutes could be attributed to an identical break point, but their perceptible differences were not due to modifiers, since the stocks were isogenic to one another. Thus the differences could be attributed to the effects of one or more genes in the bobbed and block A regions. "All in all, we do not wish to give the impression that our results with the three scutes here reported upon actually support the idea of a finely divisible gene. They do, however, give evidence for the position effect, and this phenomenon in itself connotes that the gene, as defined by the mutation-allelism test, extends over a larger region than that defined by breakage, crossing over, or self-reproducibility, and that, in fact, the regions of successive genes, as defined by the mutation-allelism test, do not merely adjoin but overlap each other. On the other hand, our present results do not, in themselves, give evidence of a finer divisibility than that found in our more general study of breakage in the scute region. . . . As we have seen, they can be interpreted in either way with roughly equal plausibility. . . . Hence, so far as our present results are concerned, this matter must still be left an open question."[267]

Muller and Raffel's paper was written in 1938, but its publication was

delayed until 1940 because of the outbreak of World War II. Muller had
left the U.S.S.R. on the advice of his colleagues when the genetics controversy became too "political" and the proponents of formal genetics began
to yield to the "Lysenkoist" groups. Muller managed to obtain an exit permit as a "volunteer" in the Spanish Civil War, and from there he emigrated
to Great Britain, where he was appointed to the staff of the University of
Edinburgh. In 1940, Muller reconsidered the implications of the work he
had done with Prokofeyeva, Ellengorn, and Raffel in the mid-1930's. "One
line of evidence indicating a segmental structure of some sort in the chromonema is provided by my findings of the restricted number of the genetically distinguishable positions of breakage producible by X-rays in the neighborhood of the loci of scute and yellow in Drosophila. . . . The only plausible
escape from the conclusion that there is a rather coarse segmental structure
here is to suppose that between these 'genes'—or minute chromosomal
regions whose functions we are observing—there lie longer stretches of genetic material, having so little or so unimportant functions that their total
loss would still leave the organism viable and normal appearing. In that
case the visible 'genes' would give the appearance of being discontinuous
only because they were separated by long interspaces of relatively inert genic
material and so the discontinuity would in a sense be of a functional nature
only."[241]

It is somewhat surprising that the intensity of activity in the mid-1930's
on the gene problem diminished so rapidly and that interest in it did not
return until the 1950's. Partly, World War II had disrupted international
co-operation in science; partly a new generation of participants, with different outlooks, was exploring new problems and new organisms with new
techniques. The "left-right" test has not been used since then for further
analysis of scute or other regions of Drosophila chromosomes nor has it been
applied to the analysis of other organisms. The number of cases of scute
that could be analyzed by the left-right test increased by three in these fifteen years, providing sixteen cases. Thus, in 1955, with the posthumous
and somewhat pessimistic statement of Stadler on the gene, and with the
growing belief of a new generation that little or no evidence existed for
structural discontinuity of the chromosome, Muller reviewed the problems
of continuity and gene mutation in a paper, "On the Relation between
Chromosome Changes and Gene Mutations."

"There are two basic questions, both as old as genetics but still warmly
disputed. First, have we a right to speak of individual genes as separable
bodies rather than only as convenient mental isolates, conceptually cut out
of an uninterrupted genetic material, or chromosome, of dimensions larger
than they? Second, if the expression 'the individual gene' does correspond
to a material reality of about the order of magnitude which it has been
conceived to have, do we have a right to regard any, or many, of the
Mendelian differences with which we deal in genetics as representing
changes, chemical in nature, within these individual genes; or are these

differences caused only by decreases and increases in number and possibly size (i.e., in number of identical parts), and by changes in mutual arrangement, of genes that themselves remain essentially unchanged, or even unchangeable?"[245] The analysis of sixteen scute rearrangements, however, argued against the structural continuity of the chromosome in *Drosophila*. "The striking fact is that all these 16 breaks fell into the limited number of positions that we have noted. There was no suggestion of any positions intermediate between them. For, when recombinants were produced involving any two cases having breaks attributed to the same position, both complementary classes of recombinants appeared to be as viable, fertile, and on the whole as phenotypically normal as that unrecombined original type whose positionally 'mutant' locus they carried. . . . In this connection it is of interest to note that at the time our paper of 1934 was published, when only eight breaks had yet been located with reference to each other, all four of the above positions of breakage had already been ascertained. That is, in the twenty years since that time, during which another eight breaks have been located genetically, no further positions of breakage have been found."[245]

A recalculation of the size of the gene, based on the scute-19 deletion and a better knowledge of the total length and number of strands of the salivary chromosomes, "allowed us to derive a maximum estimate for gene size in Drosophila of 1/30 micron, cubed, and a minimum for gene number of about 10,000."[245] No further evidence of a positive nature could be given to show how many of the operational criteria for defining a gene coincided. "It seemed very likely, both *a priori* and on the basis of the scant evidence that exists in the scute region that the breakage gene and the crossover gene would coincide, but definitive tests of this point have been lacking. On the other hand we already know that the regions defined by the phenotypic expressions or allelism are often larger."[245]

Since 1934, then, no further progress of a novel nature has been revealed by the application of the "left-right" test and cytology to the scute locus. The doubling of analyzed rearrangements in that period of time, without the addition of new break points, remains the only positive evidence that Muller and his colleagues have provided for chromosomal continuity, and it would be a surprising revelation if those four break points were not associated with structural discontinuity of the genetic material.

CHAPTER 18

The Target Theory: A Successful Failure

Occasionally a sound idea applied to one system is immensely successful, but when applied to another system, apparently similar in its characteristics, it turns out to be a failure. This is commonly experienced in physiological, pharmaceutical, and medical research in which the differences between one batch of animals and a second are often found to be profound. It is surprising, however, when the principle being studied is a physical principle and the investigators are physicists. For in these situations, the uniformity of nature expected by "naïve faith," as Whitehead once described this scientific attitude, is unexpectedly not found. Such was the history of the target theory.

The target theory may be illustrated with a simple example. If, in a parlor game, I were blindfolded and placed in a room of known size filled with a certain number of balloons, I could make an estimate of the size of the balloons if I knew how many were in the room (or the number, if I were told the size of the balloon). To do this I would only require an air pistol, firing it at random in the room, and the number of balloons popped would be a basis for making the estimate. Also, I would have to calculate the number of shots fired per hit. Obviously, the more balloons in the room, or the larger the balloons, the more likely would it be that a hit would occur.

This principle was first applied to a biological problem by J. A. Crowther in 1924. He had read a paper the year before by Strangeways and Oakley, who had found that cells of chick embryos growing *in vitro* were killed by x-rays in a predictable exponential progression. "It seemed interesting to consider whether the probability might not be due to the x-rays themselves and represent the probability that a given structure in the cell would actually be affected by the incident radiation."[92] Crowther used the number of ionizations produced by x-rays per roentgen per cubic centimeter of air and the number of atoms per cubic centimeter of air as the basis for his estimate. By calculating the time required for killing half the cells with x-ray treatment, he was able to compute the "sensitive volume" of the inferred object which was apparently hit by the ionization. This estimate, about 1/2500 mm. was considerably smaller than the diameter of the cell, which was about 1/100 mm. Such an assumed spherical object might, of course, have no physical reality at all, and only represent an abstraction. Crowther, however, thought otherwise. "It is, for example, of the same order of magnitude as the centrosome, a body which is supposed to play an important part in the phenomenon of mitosis."[92]

After the introduction of x-radiation for mutagenesis in *Drosophila* and other organisms, a number of attempts were made to use the target theory for estimating the size of the gene. Actually, Muller and Mott-Smith had made such an estimate in 1930, but they rejected this approach because the assumptions required were inconsistent with the biological properties of the mutation process. In this unpublished attempt, the mutation frequency from red to white was about 1 per 1000 sperm at 5000 r. Thus, about 5 million r would induce a mutation to white in each sperm. The number of ionizations per cubic centimeter of "protoplasm" was estimated as 10^{19}. From this, the supposed volume of the gene would have a diameter of 0.01 μ and it would contain about 2700 atoms,* "assuming that the sensitive volume is of atomic make up like that in protoplasm."[241] This was a considerably smaller size of the gene than had been estimated, at that time, by either the Poisson method, the crossover method, or the cytological method.

While Muller and Mott-Smith in 1930 rejected their estimate of the size of the gene based on the target theory, O. Blackwood independently suggested the same method for calculating the size of the gene or its sensitive volume for mutation. "The diameter of these bodies in the cells of the fruit fly . . . is roughly estimated to be 600A. The writer has computed the approximate number of ions produced in such a gene when exposed to x-rays of known intensity for a known time. Using Patterson's experimental value for the percentage of exposed flies showing a certain mutation, about one per cent of the atoms in a gene are found to be 'sensitive' (i.e. their ionization is assumed to be accompanied by mutation). If the sensitive material were concentrated in a spherical nucleus, its diameter would be about one-fifth that of the gene."[43] This estimate of 130A was changed to 200A the following year by Blackwood when he corrected for the x-ray absorption in

* This would correspond to a strip of DNA capable of coding a polypeptide sequence of ten amino acids.

the system he used. "In causing mutation by X-rays only one ion pair is produced if the radius is 130A and thirty if Muller's maximum value (600A) is correct."[44]

While Blackwood was cautious not to commit himself to the identification of the target volume with the gene, he did favor this assumption, as the title of his paper implied: "X-ray evidence as to the size of a gene." This implication was accepted by J. W. Gowen and H. Gay in a similar attempt to calculate the number and sizes of genes in *Drosophila*. "X-rays are absorbed in units. An absorbed unit has a sphere of effect within which it acts. It is this purely physical quantity which is being measured. The result is biologically significant in that it sets the upper limit of size of these elements whose alteration produces the observed results. The upper limit for the size of the gene is 1×10^{-18} cm^3. This value is below microscopic vision and therefore considerably less than the chromomeres assumed by Belling to represent genes."[156] Gowen's estimate of 10^{-18} cm^3 is approximately that of a sphere with a diameter of $1/50$ μ or about 200A.

In 1935 Max Delbrück, then a physicist, collaborated with N. W. Timofeef-Ressovsky and K. G. Zimmer in a physical analysis of the mutation process and the structure of the gene. Using the principles employed by Crowther, Blackwood, and Gowen, they computed an even smaller volume for the target than did their predecessors. This "sensitive volume" was also based on a study of the white locus. "To carry out this experiment we employed radiation treatment on a fairly frequently occurring mutation, namely the mutation from normal to eosin (w^+ to w^e), in which an average of one mutation in a little less than 7000 gametes occurs at a dose of 6000 r. Since the mutation rate is proportional to dose, we can estimate that the total dose of $6000 \times 7000 = 42,000,000$ r has a probability of inducing just one mutation. On the other hand, about 2×10^9 ion pairs are produced in a c.c. of air under standard conditions, or, in water or organic material, about a thousandfold more; thus about 2×10^{12} ion pairs are produced altogether. Multiplying this by the dose, 42×10^6 r, gives about 10^{20} ion pairs, each with an energy of about 30 ev. Since about 10^{23} atoms are present in a c.c. of this treated material, we have ionized about one one-thousandth of these atoms. The fact that the reaction occurs with a probability of inducing only one mutation at this dose, must mean that the dissipation of the energy does not occur with a maximum velocity."[309]

The energy of the ionization, 30 ev, was enormous compared to the energy calculated by Delbrück to induce a mutation in a gene. This calculation revealed a surprising relation to the biologist, who is usually not familiar with quantum mechanics. Very small differences in the threshold energy produced profound differences in the stability of molecules. Thus for a threshold energy of 0.9 ev, a gene would not be stable, mutating at a frequency of once every tenth of a second. For a threshold energy of 1.5 ev, the gene would mutate once every sixteen months; and for a threshold energy of 1.8 ev this stability would be raised to 30,000 years! The implications of this for the effect of temperature on mutation was appreciated by Muller. "On the view of gene mutation as a microchemical accident it was to have been expected that within the range of temperature normal to the organism . . . a rise of 10° C would increase the rate of change . . . several times, according to van t'Hoff's rule. . . . In connection with the effect within the range of temperature normal to the organism, the physicist Delbrück, collaborating with Timofeef-Ressovsky, has made a very important contribution. He points out that, as has been well known to physical chemists, the amount of increase in the rate of a reaction, caused by a rise of temperature, depends, according to a known formula, upon the rate of the reaction itself, at any one given temperature. If now, we accept the previously presented view of the ordinary mutations as being due to microchemical accidents in the same sense as the changes of molecules in other chemical reactions are, and if we then take into account the rate of the mutation reaction of the gene at a given temperature, as determined by the data on mutation frequency, we find that, corresponding with the extremely slow rate of this reaction (many thousands of years intervening between one change of an individual gene molecule and the next), its rate of rise with temperature should be exceedingly high, as compared with the rate of rise of ordinary reactions. The expected 'Q_{10}' co-efficient thus calculated turns out, in fact, to have the unusual value of 6 to 8, instead of the ordinary value (which lies between 2 and 3). . . . Re-examining my own and Altenburg's earlier results, derived from three separate sets of experiments, I find that all these agree in giving this high value of Q_{10}, when the time-factor is properly taken into account. . . . These correspondences of the observed and calculated values of the Q_{10} . . . constitute a striking confirmation of the idea of gene mutation as being a result of random inter- and intra-molecular motions."[240]

In 1940 Muller repudiated the major implications of the target theory either for measuring the size of the gene or for estimating the critical "sensitive volume" which determined its mutation. The calculation that he had made with Mott-Smith (about 2700 atoms contained in a sphere having a diameter of 0.01μ)' as well as the estimates of Blackwood, Gowen, and Delbrück, were virtually meaningless. "It will be seen that this identification involves the following main assumptions: (1) that any ionization (or other activation?) of any atom within a gene unfailingly results in a genetic change; never being ineffective or followed by restitution, (2) that this genetic change is always a genetic change of the type being looked for and detectable by the method used, (3) that the ionization of no atom outside the gene can, through a transmission of energy, result in such a change of the gene. Unfortunately for the validity of the method as a means of finding gene or chromatin size, every one of these three assumptions is not merely gratuitous but improbable on theoretical grounds and there is in fact strong empirical evidence against each of them, in connexion with either breakages or gene mutations or both."[241]

One of the reasons that Muller, and Timofeef-Ressovsky before that, had not equated the sensitive volume with the gene itself was the independence of the mutation rate from the wavelength of the ionizing radiation. The number of mutations produced was proportional to the number of ionizations formed whether those ionizations were dense or sparse. Using the target theory, let us assume that we have a large number of

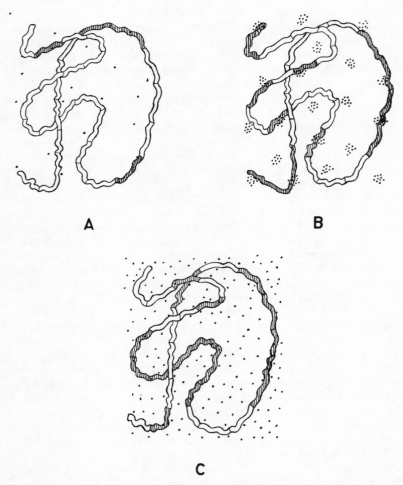

A   B

C

Figure 26. The Target Theory. In A, twenty ionizations are distributed at random near a chromosome; four genes are hit. In B, the dosage is increased tenfold but the distribution of the 200 ionizations is clustered into twenty groups of ten ionizations. Twelve genes are hit. In C, the 200 ionizations are randomly distributed, resulting in fourteen hits. If the effects of ionization were indirect (via chemical routes), then there would be no difference between B and C.

bodies exposed to a randomly distributed batch of ten ionizations such that two of the ionizations score hits in the objects. If, by changing the wavelength, we increased the ionizations tenfold, but the density was clustered around the same points that were present in the sparse situation, then there would still be only two hits, since it would be unlikely that such localized densities would have any effect outside of their sensitive volumes. However, if the mutations were indirectly produced, rather than through direct hits in a target, then the mutation frequency would be higher; ten times higher, in fact, since the mutation frequency would be proportional to the number of ionizations. Such considerations "lead us to infer . . . that (1) only a comparatively rare ionization, even when it occurs within a gene, leads to a mutation, and especially to a detectable one, and (2) ionizations outside the gene also can produce genetic changes, certainly those of the nature of breaks, and perhaps also gene mutations. The 'sensitive volume' is thereby relegated to an expression for the number of atoms which at any given moment are so situated, and in such a condition as regards their configuration and dynamics, that their ionization, if produced according to some randomly specified pattern out of the various possible energy patterns for an electron 'hit' will result in a mutation of the specific kind. . . . Thus a more intelligible picture would be that a 'potential sensitive volume' comprises the whole region within which an ionization might on one occasion or another result in a mutation, and to say that the 'sensitive volume' . . . consists of this 'potential sensitive volume' divided by the average chance that any atom in the latter would, if ionized, give rise to the mutation."[241] Such a modification eliminated the initial promise of the target theory: the estimation of the size of the gene itself. "Because of all these sources of error, any apparent agreements between the 'sensitive volumes' hitherto found and the maximum size of the gene as estimated by quite different methods must be regarded as purely fortuitous."[241]

Despite these criticisms, the target theory was defended by D. E. Lea during the 1940's. In his book, *Actions of Radiations on Living Cells*, Lea maintains that the various objections were either negligible or could be interpreted in a different way. Thus "if we have reason to believe that a particular action of radiation is of the single-ionization type, and yet find that when we work out the size of the target from experiments made with different radiations, we fail to get consistent results, but the estimates increase in the order of increasing ion-density, then a possible explanation is that the target is filamentous. . . ."[177] Lea rejected the assumption of indirect effects of the ionization chiefly because the direct effects were consistent with the results obtained in the inactivation curves of viruses and other systems. Using soft x-rays, neutrons, and alpha-rays, Lea obtained a range of 4 to 8 mμ for the diameter of the sensitive volume which he equated with that of the gene itself. Such a diameter, 40A, was the smallest estimate made for the size of a gene.

In the subsequent years of its history the target theory disappeared as a means of estimating gene size; it has however been useful in other ways. The number of symbiotic or infectious particles in a cell can be determined by target theory, especially when radiation is used as the basis for hitting (and killing) the target. Preer, for example, made effective use of radiation to destroy particulate bodies, kappa, which were present in the cytoplasm of some strains of *Paramecium*. The target theory prediction of kappa size was consistent with that obtained later on by direct cytological measurement.

Most of the complexities ignored or not recognized during the first years of its usage in genetics were shown to be important. Thus indirect effects of radiation were demonstrated in bacteria by Stone, Wyss, and Haas, who found, in 1952, that it was not necessary to expose the bacteria to ionizing radiation to produce mutations. If the unexposed bacteria were placed on irradiated medium, mutations would occur. This indirect route for mutagenesis was probably a consequence of peroxide formation since high catalase contents in cells (which break down the peroxides) made the cells "resistant" to the indirect mutagenesis through irradiated medium.

Lea's estimates were so low that their biological significance was questionable. "A volume $6m\mu$ in diameter is about a thousandth or less of the maximum gene size which had been calculated by Muller by several different methods. The term 'maximum' is used here to mean the size which a gene would have on the admittedly only limiting assumption that a chromosome, when in the condensed state in which it occurs at metaphase or in a spermatozoan, contained no material other than its genes. It seems likely that even when in this form the chromosome does contain some other material. However, it seems very unlikely, especially in view of the survival value of having the spermatozoan trimmed down to the smallest possible dimensions, that the gene material should occupy only a thousandth part of the chromosome's bulk. . . . Such an object, containing only a few thousand atoms of six main types and further limited by the necessity of these atoms being grouped into only a few hundred amino-acid and nucleotide blocks of a few tens of standard kinds, would hardly seem capable of developing that unlimited diversity, specialization, and nicety of functioning which genes, in their varied representatives, have attained."[244] If, in contemporary thinking, we estimated the number of nucleotide pairs in a gene determined by target theory to have a diameter of 60A, this would be about 80 pairs.

The major value of the target theory to genetics was not its estimates of the size of genes or sensitive volumes within the cell, but its stimulation of biophysical approaches to genetics. The interests of a few physicists were aroused by genetic phenomena; Delbrück found far more biological problems worth exploring than the size of the gene estimated by target theory; the appeal of biology was too powerful to resist and Delbrück's healthy injections of the physicist's outlook led eventually to a school of molecular genetics. So too, Erwin Schrödinger took time out from his interest in quantum mechanics to think of its implications for biology. Schrödinger was captivated by Delbrück's model of mutagenesis, and he admired the use of the target theory as a basis for estimating the size of the gene. Schrödinger's popularization of these accounts in his book *What is Life?* stimulated other physicists to consider ways to approach biological problems as molecular problems. Among those stimulated by Schrödinger's account of target theory and "quantum genetics" and the existence of a genetic "code-script" was Seymour Benzer, whose contributions to the gene concept are among the most exciting adventures in the history of genetics.

CHAPTER 19

A Prophecy Fulfilled:
The One Gene:One
Enzyme Hypothesis

Shortly after the rediscovery of Mendelism, a series of papers appeared in medical journals, authored by Archibald E. Garrod. Garrod was a physician whose interests extended to biochemistry and pathology. It had been noted for some time that defects of human metabolism were frequently detectable through urine analysis. This was particularly true for diabetes. But quite different from such metabolic diseases "is the course of the anomalies of which I propose to treat . . . and which may be classed together as inborn errors of metabolism. Some of them are certainly, and all of them are probably, present from birth. The chemical error pursues an even course and shows no tendency to become aggravated as time goes on. With one exception they bring in their train no serious morbid effects, do not call for treatment, and are little likely to be influenced by any therapeutic measures at our disposal. Yet they are characterized by wide departures from the normal of the species far more conspicuous than any ordinary individual variations, and one is tempted to regard them as metabolic sports, the chemical analogies of structural deformities."[133]

One of these "metabolic sports" was a visible disorder, albinism; the rest were physiological defects detected primarily through the urine. "Taking all the known facts into consideration, the theory that what the albino lacks is the power of forming melanin which is normally possessed by cer-

166

tain specialized cells is that which has the most favour and is probably the true one. If so, an intracellular enzyme is probably wanting in the subjects of this anomaly, an explanation which, as we shall see later, brings albinism into line with some other inborn metabolic errors, of which a similar explanation is at least a possible one."[133]

Among the physiological defects cited by Garrod was the condition called alkaptonuria. Individuals with this disease had a defect which might be embarrassing on rare occasions, but it would not be injurious to their health. On exposure to air, the urine of patients with alkaptonuria turned various shades of red, brown, or black. In many families the condition was noticeable at birth from the stained discoloration of diapers. The presence of discolored urine also made its victims poor insurance risks, although physically they appeared to suffer no ill effects. A congenital defect, however, is not necessarily an inherited defect. The suspicion that the defect was produced at conception arose in 1902 when Garrod examined the marriage records of families having alkaptonuric children. A larger than to be expected number were first cousin marriages. But "it appears to me that the strongest argument which can be adduced in favour of this view that alkaptonuria is a Mendelian recessive is afforded by the fact that albinism, which so closely resembles it in its mode of incidence in man, behaves as a recessive character in the experimental breeding of animals."[133]

The actual metabolic defect of alkaptonuria was shown by Garrod to be an inability to utilize the benzene ring from the amino acids, tyrosine and phenylalanine. "We may further conceive that the splitting of the benzene ring in normal metabolism is the work of a special enzyme, that in congenital alkaptonuria this enzyme is wanting, whilst in disease its working may be partially or even completely inhibited."[133]

Although Garrod attempted to popularize his theory of "inborn errors of metabolism" in a series of Croonian lectures to the Royal Society, he failed to generate a close co-operation between biochemists, pathologists, and geneticists. From time to time his work would be cited, but some thirty-five years were required before Garrod's theory was revived and hailed as a major contribution to genetics.

Bateson, for example, not only inferred the possibility of recessive lethal conditions as a consequence of Garrod's discussion of alkaptonuria, he acknowledged that the factors of heredity were associated with the production of enzymes, a view pursued by his student Onslow (née Wheldale) in the genetics and biochemistry of flower pigmentation. Bateson lacked the biochemical training to extend this view, and he failed to carry the message to the biochemists themselves. Furthermore, the nature of these factors, in 1907 when he presented his Silliman lectures on *Problems of Genetics*, was unknown. "We must not lose sight of the fact that though the factors operate by the production of enzymes, of bodies on which these enzymes can act, and of intermediary substances necessary to complete the enzyme action, yet these bodies themselves can scarcely be genetic factors, but consequences of their existence. What are the factors themselves? Whence do they come? How do they become integral parts of the organism? Whence, for example, came the power which is present in a White Leghorn of destroying—probably reducing—the pigment in its feather?"[20]

A daring attempt to bring biochemistry into genetics was made in 1917 by L. T. Troland. Troland was stimulated by the novel discoveries of atomic structure and the philosophic implications of radioactivity and other developments in physics. "It is perhaps not surprising that the astonishing progress of general physics during recent times should thus far have failed to exert any very notable influence upon the science of biology. From the point of view of the physicist, biological problems must be regarded as questions of special material structure, usually of a very intricate character, and involving the arrangement and history of units of matter for the most part larger than those upon which attention is immediately concentrated. . . . However, a critic who sees current events in the light of the history of science can hardly escape a twinge of disappointment at the recrudescence in biological theory, at the present time, of the doctrine of vitalism. The present . . . is an hour of triumph of the monistic theory of nature, and yet now, more frequently than during the nineteenth century, men eminent in biology seem to quail before the complexity and delicacy of the life processes, and, while uttering mechanistic truths about life, to offer them as sacrifices to a spirit of vagueness and discouragement. . . . It is my belief that this rejuvenation of mysticism and Aristotelian teleology is due . . . to their neglect of modern physics and of the methods of thought pursued in that science."[310] Troland agreed with Bateson that the demand "is not for new biological facts, but for physico-chemical conceptions in terms of which a chaos of biological facts, already at hand, can be explained, or systematized."[310]

Troland's article was not based on experimental evidence. He specifically acknowledged that his purpose was polemical. There was one theory which he hoped would stimulate a "physical" outlook in biology. "It has for some years been my contention that the conception of enzyme action, or of specific catalysis, provides a definite, general solution for all of the fundamental biological enigmas: the mysteries of the origin of living matter, of the source of variations, of the mechanism of heredity and ontogeny, and of general organic regulation. In this conception I believe we can find a single, synthetic answer to many, if not all, of the broad, outstanding problems of theoretical biology. It is an answer, moreover, which links these great biological phenomena directly with molecular physics, and perfects the unity not only of biology, but of the whole system of physical science, by suggesting that what we call life is fundamentally a product of catalytic laws acting throughout the long periods of geologic time."[310] The enzymes, in Troland's view, extended beyond the products of genes themselves. "The suggestion that the germ-cell contains 'determiners' for the production of enzymes, which, in turn, regulate certain aspects of the development, is a common one. . . . On the supposition that the actual Mendelian factors are enzymes, nearly all . . . general difficulties instantly vanish, and I am not acquainted with any evidence which is inconsistent with this supposition."[310]

Although Troland's views were incorporated by Muller in his discussion of "Variation due to change in the individual gene," the assertion that genes are enzymes was rejected. "The chemical composition of the genes, and the formulae of their reactions, remain as yet unknown. We do know, for example, that in certain cases a given pair of genes will determine the existence of a particular enzyme (concerned in pigment production), that another pair of genes will determine whether or not a certain agglutinin shall exist in the blood, a third pair will determine whether homogentisic acid is secreted into the urine ('alkaptonuria'), and so forth. But it would be absurd, in the third case, to conclude that on this account the gene itself consists of homogentisic acid, or any related substance, and it would be similarly absurd, therefore, to regard cases of the former kind as giving any evidence that the gene *is* an enzyme, or an agglutinin-like body. The reactions whereby the genes produce their ultimate effects are too complex for such an inference."[225]

The revival of an interest in the biochemistry of gene function developed rapidly through the efforts of George W. Beadle.* Originally trained as a maize geneticist, Beadle took postdoctoral training at the California Institute of Technology where Morgan and Sturtevant attempted to develop a Division of Biology which abolished the traditional departmental separation of zoology, botany, and microbiology. Working with Sturtevant, Beadle became interested in the development of eye pigmentation in *Drosophila*. He traveled to Paris to work with Boris Ephrussi on the techniques of transplanting the embryological rudiments of eyes into new larval hosts. Their success in working this out enabled them to make use of the large number of different eye color mutants in *Drosophila*. They were rapidly rewarded in the use of their technique. "Sturtevant has shown that vermilion eye color is, under certain conditions, not autonomous in its development in mosaics. In transplants it is likewise not autonomous, a vermilion (v) eye implanted in a wild type host develops the pigmentation characteristic of wild type. By means of transplantation we have been able to study many combinations not easily obtained in natural mosaics and in this way have found that cinnabar (cn), an eye color phenotypically similar to vermilion, is not autonomous in its pigment differentiation. Two other

---

* J. B. S. Haldane should also be given credit for this interest in biochemical genetics because of his effectiveness as a theoretician. His popularization of Garrod's work preceded Beadle and Tatum's references to "inborn errors." Haldane also sought to relate genetics and biochemistry in a monograph published on this theme. Unfortunately, Haldane seldom carried his theories to the laboratory, and the contributions to biochemical genetics in the 1940's are properly attributed to Beadle and Tatum.

eye color mutants, scarlet (st) and cardinal (cd) likewise phenotypically similar to vermilion, are, however, completely autonomous in their pigment development in all the combinations in which we have studied them."[31] The extension of this technique to reciprocal implantations yielded some unexpected information. "We have found that a v disc in a cn host gives a wild type eye, but that a cn disc in a v host gives a cn eye. . . . v and cn implants behave in the same way . . . in a claret (ca) host, both v and cn implants are autonomous; in st or cd hosts they are both modified in wild type. This corroborates the conclusion drawn from reciprocal transplants between v and cn in indicating that the v and cn host-implant influences are genetically—and presumably chemically—closely related."[31]

A year later, Beadle and Ephrussi discussed the significance of their technique and their findings. "Prominent among the problems confronting present day geneticists are those concerning the nature of the action of specific genes—when, where, and by what mechanisms are they active in developmental processes? Despite the recognized importance of such questions as these, relatively little has been done toward answering them, a situation not at all surprising considering the difficulty of getting at these problems experimentally. Even so, promising beginnings are being made; from the gene end by the methods of genetics, and from the character end by biochemical methods. Probably the one factor which has played the most significant role in retarding progress in this field is the fact that relatively little is known from a developmental point of view about those organisms that have been studied most thoroughly from the genetic point of view, and, on the other hand, little is known genetically in those organisms that have been most studied from the developmental point of view."[32]

The non-autonomous behavior of claret, vermilion, and cinnabar under certain conditions required some hypothesis to relate them, if possible, and in so doing to explain how the mutants produced their effects in transplantation studies. "A simple, and, it seems to us, plausible, hypothesis may be of help in answering these questions. Such an hypothesis assumes that the ca^+, v^+, and cn^+ substances are successive products in a chain reaction. The relations of these substances can be indicated in a simple diagrammatic way as follows: $\rightarrow ca^+$ substance $\rightarrow v^+$ substance $\rightarrow cn^+$ substance. In such a scheme, we assume that . . . the mutant gene ca in some way produces a change such that the chain of reactions is interrupted at some point prior to the formation of ca^+ substance; hence a ca fly lacks ca^+, v^+, and cn^+ substances. . . . The mutant gene cn stops a reaction essential for the change of v^+ substance to cn^+ substance; hence a cn fly lacks cn^+ substance but has the ca^+ and v^+ substances."[32]

In the hopes of achieving a biochemical basis for this hypothetical developmental scheme, Beadle worked with the biochemist E. L. Tatum. They suspected and were able to demonstrate that the brown eye pigment lacking in the various non-autonomous eye color mutants was associated with the conversion of the amino acid tryptophane into kynurenin. "In an attempt to summarize and indicate the interrelations of the facts . . . we have drawn up a partly hypothetical scheme. . . . In this, tryptophane is shown as the precursor of kynurenin (v^+ hormone), with the transformation under the control of the vermilion gene. Kynurenin may be transformed either to kynurenic acid, which is inactive, or to cn^+ hormone which is in the chain of reactions leading to brown pigment. . . . We suggest that the suppressor of vermilion gene may control the oxidation of kynurenin to kynurenic acid. In the presence of the recessive alleles of this gene in an otherwise vermilion fly, kynurenin is restored because the competing reaction leading to kynurenic acid is blocked."[33] Each of the known biochemical steps in the pathway towards brown pigment seemed to be the work of an individual gene. "Since a fly in which the ultimate recessive alleles of the white gene are present lacks both red and brown pigments, we can infer that in the development of these two pigments there is a common step. This inference is based on our faith in the unproved assumption that a given gene has a single primary action."[33]

This relation of one gene to one primary action took on more defined form when the nature of the primary action was suggested to be enzyme production. "Throughout the scheme we have indicated genes acting through the intermediation of enzymes. In a sense this is a purely gratuitous assumption, for we have no direct knowledge of the enzyme systems involved. Since, however, we know that in any such system of biological reactions, enzymes must be concerned in the catalysis of the various steps, and since we are convinced by the accumulating evidence that the specificity of genes is of approximately the same order as that of enzymes, we are strongly biased in favor of the assumption. In this we make no claim to originality, for it has many times been suggested by geneticists that there may be a close relation between genes and enzymes. It is, of course, possible that the immediate products of many genes may be enzymes or their protein components. At the present time, however, the facts at our disposal do not justify the elaboration of hypotheses built on this assumption."[33]

Later in that same year, 1941, Beadle and Tatum developed a means to test their tentative hypothesis of one gene : one primary character : one enzyme. In their studies of the eye pigment pathway Beadle and Tatum were limited to only a few mutants suitable for testing. Furthermore they had no way of applying a biochemical approach to other mutant systems, such as bristle formation or wing shape. For this reason they made use of another organism, brought by Morgan from the laboratory of Dodge and passed on to Carl Lindegren for genetic studies. This organism, Neurospora, was a fungus. "Considerations . . . have led us to investigate the general problem of the genetic control of developmental and metabolic reactions by reversing the ordinary procedure and, instead of attempting to work out the chemical bases of known genetic characters, to set out to determine if and how genes control known biochemical reactions. The ascomycete Neurospora offers many advantages for such an approach and is well suited to genetic studies. Accordingly, our program has been built around this organism. The procedure is based on the assumption that x-ray treatment will induce mutations in genes concerned with the control of known specific chemical reactions. If the organism must be able to carry out a certain chemical reaction to survive on a given medium, a mutant unable to do this will obviously be lethal on this medium. Such a mutant can be maintained and studied, however, if it will grow on a medium to which has been added the essential product of the genetically blocked reaction."[34] In their first experiment Beadle and Tatum found three biochemical mutants, all affecting vitamins, that required an appropriate supplementation of the missing vitamin in order to survive. Since the synthesis of vitamins was known to be under enzymatic control, Beadle and Tatum concluded that "it is entirely tenable to suppose that these genes, which are themselves a part of the system, control or regulate specific reactions in the system either by acting directly as enzymes or by determining the specificities of enzymes."[34]

By 1945 Beadle and Tatum had detected numerous biochemical mutants affecting large numbers of products, establishing the generality of their technique for the study of biochemical pathways. "When metabolically deficient strains are investigated . . . the great majority of them prove to differ from the original wild-type in the alteration of one single gene. Occasionally double mutants are obtained that require two substances for growth. In all but one instance, these have proved to have two mutant genes. Their frequency is of the order of magnitude expected on the assumption of independent occurrence of the two mutations. . . . The one exceptional case so far encountered involves a mutant strain which requires both valine and isoleucine for normal growth. It is supposed that the related amino acids have a common step in their synthesis and that it is this step which the gene in question normally controls."[30] The success of the direct study of biochemical genetics through Neurospora was a major achievement. It brought into being the union of genetics, biochemistry, and developmental biology. Within a few years Joshua Lederberg would provide the essential tools for applying biochemical genetics to bacteria, bringing in the microbiologists. Bacterial genetics, in turn, would draw in the physicists who, through Delbrück's initiative, were using bacteriophage (d'Herelle bodies) as genetic systems in the hope of establishing molecular models of genetics.

The combination of these different approaches led to a co-operation of genetics and biochemistry in the late 1940's and early 1950's. The recognition that this co-operation was coming was emphasized by Beadle and Tatum. "It is perhaps unnecessary to state the obvious conclusion that if one is to understand the metabolism of the organism in the most complete way possible, genes must be taken into account. Too often in the past these units have been regarded as the exclusive property of the geneticist. The biochemist cannot understand what goes on chemically in the organism without considering genes anymore than a geneticist can fully appreciate the gene without taking into account what it is and what it does. It is a most unfortunate consequence of human limitations and the inflexible organization of our institutions of higher learning that investigators tend to be forced into laboratories with such labels as 'biochemistry' or 'genetics.' The gene does not recognize the distinction—we should at least minimize it."[30] Beadle and Tatum's work thus made a reality of Garrod's hope, expressed in 1908, that someday biochemistry and pathology would be two aspects of the same problem and Muller's hope, in 1921, that genetics, microbiology, chemistry, and physics would eventually participate in a common problem: the nature of the gene.

CHAPTER 20

The Plasmagene Theory

A cell is an exceptionally complicated system with numerous macromolecules organized to carry out metabolism. The nucleus, in most cells, constitutes a minor fraction of the total cellular volume. Why then was the nucleus and not the cytoplasm chosen as the probable site for the units of heredity? At the close of the nineteenth century the German cytologists, particularly T. Boveri, pointed out the inequality in size between eggs and sperm. The elimination of most of the cytoplasm from the cells which form the spermatozoa led them to conclude that the equivalent parental contributions to heredity must be conveyed by the nuclei of the germ cells. The subsequent chromosomal mechanics of Mendelism re-enforced this view, and through the Drosophila group, the gene concept found itself firmly rooted in the chromosome.

From time to time, however, discoveries were made which indicated a non-Mendelian inheritance. This was true for certain patterns of coiling in snail shells and for numerous cases of striped or variegated leaf patterns in plants. Where the transmission of such traits was associated with the eggs rather than the sperm, the suspicion of a cytoplasmic influence was strong. Yet, despite the acknowledged cases of cytoplasmic inheritance, the general feeling among geneticists was directed towards the nucleus as the exclusive, or nearly exclusive, bearer of the genes. In the case of plants, variegation was attributed to mutations in the chloroplasts. This interpretation, first suggested by Correns, was not universally accepted, and as an alternative, the cytoplasm itself was considered to be defective, causing the loss of normal function in the plastids.

The genetic control of such a variegated condition was discovered by Marcus M. Rhoades in 1943. A recessive mutation resulted in the partial

174

or complete loss of chlorophyll from the plastids of some cells. This trait, called iojap (ij), did not give identical genetic patterns when reciprocal crosses were made. Pollen from iojap plants which fertilized ovules from normal green plants would give green progeny in the F_1. But normal pollen fertilizing iojap ovules gave progeny that were either green, variegated, or white. "A cytological examination of the white regions of *ij ij* plants disclosed that the plastids of the mesophyll cells not only lacked chlorophyll but also were much smaller than were the plastids of the normal green areas. It appears, therefore, that the *ij* gene is able to induce a modification in the plastid."[268] When heterozygous plants were formed from ovules produced by an iojap plant, the genotype would be at variance with the trait expressed. Nevertheless, "white areas of *Ij ij* plants . . . showed them to have a type of plastid similar to that found in the white areas of the parent *ij ij* plant."[268] Even homozygous plants lacking the iojap gene would be variegated if the normal ovule came from a plant showing iojap. "It is a pertinent fact that in *Ij Ij* cells the mutant plastid continues to give rise to mutant plastids; there is no control by nuclear factors on the type of plastid. Although induced by a nuclear factor, the *ij* gene, the mutated plastid, like a Frankenstein monster, is no longer under the control of its master."[268]

The genic initiation of cytoplasmic inheritance and the subsequent loss of nuclear control provided a model for studying differentiation in tissues. "The view that cellular differentiation is cytoplasmic seems to require that the cytoplasm contains elements of a hypothetical nature which are modified by interaction with nuclear products. In the case reported here a known constituent of the cytoplasm, the plastid, has been modified by a nuclear factor and is transmitted thereafter by cytoplasmic heredity."[268]

A few years before the appearance of Rhoades's analysis of the iojap gene, another organism was proving to be of complex genetic constitution despite its apparently lowly position on the evolutionary scale. Tracy M. Sonneborn had been using *Paramecium* for a number of years in the late 1930's. This protozoan, a ciliated single-celled organism, was immensely bigger than typical cells. It was visible as a fleck to the unaided eye, and there was no difficulty seeing its numerous organelles with the compound microscope. At one time these organisms were considered to be sexless, multiplying their numbers by a simple binary fission or mitosis. This proved to be false. In 1937 Sonneborn proved the existence of sexuality in *Paramecium*. "Paramecium has long appeared an outstanding example of the absence of sexual differentiation in individuals that nevertheless conjugate. In a certain race of *Paramecium aurelia*, however, I have discovered that there is functional sex diversity. Two classes of individuals exist. Members of different classes unite for conjugation; members of the same class do not."[279] Unlike the typical mating of higher organisms, the two cells were identical in appearance and could only be distinguished by their physiological response. The two cells, when compatible, joined together along a specific region of their surface, forming a conjugating pair. A small cytoplasmic bridge permitted the reciprocal exchange of nuclei. In many ways the process was analogous to meiosis and fertilization. Even the separation of germinal and somatic tissue in Weismann's theory had a counterpart in *Paramecium*. During the sexual process, "germinal" micronuclei were exchanged but the "somatic" macronuclei were not. By 1939 Sonneborn's investigations had assumed considerable significance. "May I suggest that the surest value of the new knowledge of sexuality in Paramecium lies in what it may contribute not so much to the field of sexuality as to the field of genetics proper. Here it has provided the basic technique for controlling and obtaining readily the crosses necessary for genetic analysis. Lack of this has until recently greatly impeded progress in genetics of the ciliate Protozoa. With it, genes and typical Mendelian inheritance were soon found; and by means of this, a clear demonstration of the basic nuclear processes of conjugation and autogamy. Approaches to two general genetical problems, for the study of which the ciliate Protozoa are especially favorable have already been made: the problems of the interactions of genes and cytoplasm and of the interaction of genes and environment. On these and other general problems of genetics, the prospects for important contributions from the ciliates seem excellent."[280]

Sonneborn's optimism was justified a few years later with the discovery of a new phenomenon. While similar to the iojap situation in maize and appearing in the same issue of the *Proceedings of the National Academy of Science*, Sonneborn's discovery was made in a system that provided a far more extensive analysis of nuclear and cytoplasmic inheritance. "The present paper reports a previously unknown system of relations between a gene and a cytoplasmic substance, both of which are required for the development of an hereditary character. When some of the cytoplasmic substance is present, the gene controls its continued production; but when the cytoplasmic substance is absent, the gene cannot initiate its production. Addition of the cytoplasmic substance to an organism, lacking the character dependent on it, but containing the required gene, results in the continued production of the substance, in the development of the character determined by the combined presence of gene and cytoplasmic substance, and in the hereditary maintenance of the character in successive generations."[281]

The character studied by Sonneborn was a physiological "killer" substance released by certain killer strains (stock 51) of *Paramecium aurelia*. Cells killed by this substance were designated as sensitives. These sensitives were resistant to the killing action during conjugation and meiosis; thus, they could be crossed to killer strains, and the progeny of the mating could be tested for their response to the killer trait. Using a sensitive strain from stock 47, Sonneborn found that the two F_1 exconjugants retained their parental traits. All the progeny of the F_1 killer were killers, and all the progeny of the F_1 sensitive were sensitives. This failure to obtain segregation in the F_2 made it appear to be a cytoplasmic trait which was independent of the micronuclear exchange. The hypothetical killer substance in the cytoplasm was designated kappa. Killers contained kappa and the sensitive strains lacked kappa. Another sensitive strain, stock 32, also gave F_1 exconjugants that bred true to the parental types, but the F_2 progeny from an F_1 killer segregated into killer and non-killer cells, while the F_1 sensitives gave only F_2 sensitives. From these two types of crosses, Sonneborn rejected cytoplasmic inheritance as the basis of the killer phenomenon. "Full analysis shows that the interpretation in terms of cytoplasmic inheritance in Paramecium is an illusion and untenable. This raises the question as to whether the same interpretation based on similar evidence in plants is sound. . . . In plants the fundamental observations are: (1) reciprocal crosses yield different results; (2) the differences persist through subsequent generations in the female line of descent, even when all . . . of the genes in the female parent have been replaced by genes from the male parent. . . . Reciprocal crosses in Paramecium, as in plants, yield different results."[282] The existence of two sensitive strains permitted this rejection of cytoplasmic inheritance. "If, instead of using the sensitive race 32 . . . the sensitive race 47 is employed in crosses to the same killer race 51, the same results are obtained in the F_1 generation . . . but very different results follow in subsequent generations. The F_1 killers, though they have genes from both races, show no segregation of these traits in any breeding tests. . . . The progeny that derive their cytoplasm from the killer race remain killers. Likewise, the progeny of F_1 sensitive clones that derive their cytoplasm from the sensitive race remain sensitive."[282]

Apparently the hypothetical factor, kappa, could only be maintained in the presence of a dominant gene, K. Sensitives lacked kappa whether they contained K or not. However, killers had to have both kappa and gene K. When killer cytoplasm was present in an exconjugant whose genotype became k/k, the kappa factor would disappear and the cell would become a sensitive. Sensitive cells containing the gene K could become killers only if some of the cytoplasm from a killer strain were introduced through conjugation. "This evidence makes the demonstration of the determinative influence of the cytoplasm complete. . . . This conclusion would have been unavoidable if only races 47 and 51 had been available. Only the extremely good fortune of having the race 32 for use in further analysis prevented me from falling into this error. . . . The false conclusion drawn from analyses limited to the sensitive strain 47 is a consequence of the fact that this sensitive race, like the killer race 51, possesses the killer gene K."[282] Thus, where 51 is K/K and 47 is K/K there is a constant genotype and the mating of killer by sensitive results in an apparent cytoplasmic inheritance which follows the cytoplasmic descent unless an exchange of cytoplasm also occurs. Where strain 51, K/K, conjugates with strain 32, k/k, the F_1 are heterozygous in each exconjugant but only the cell with kappa in it remains a killer. From this killer the F_2 will segregate K/K, K/k, and k/k descendants, but only those having K will maintain the killer trait. The k/k F_2 cells will become sensitives as soon as their kappa disappears.

Sonneborn saw a parallel between the genetic control of a cytoplasmic trait and the phenomenon of bacterial transformation. "The Pneumococcus situation may be compared with the killer situation in Paramecium. In the latter, transfer of a cytoplasmic substance from a killer cell to a sensitive cell containing gene K will result in the continued production of this substance and the killer phenotype which depends on it. . . . As in Paramecium, when an appropriate substance essential for the synthesis of the polysaccharide is added to a cell that lacked it, the 'gene' will determine its continued production. As in Paramecium, the gene seems to be unable to initiate its production, but can continue it when the proper substance is provided to start the gene going. The fact that any strain produces only one of these 50 polysaccharides would lead one to suppose that the same gene is involved in all 50 cases and that there are more than 50 cytoplasmic materials which the gene can act upon."[282] At the time Sonneborn wrote this it was still not known that transformation was caused by a nucleic acid (DNA); the transforming principle, if similar to the kappa analogy, would be a cytoplasmic factor subject to the maintenance of a gene present in the recipient bacterial cells. The killer story also provided a new way to interpret mutation. "According to prevailing views, the physical basis of a mutation is either a change in the gene or in the number and arrangement of the chromosomes and their parts. The hereditary change from sensitive to killer, however, can occur without any change in the gene. . . . Observations on Paramecium show that mutations, in the sense of hereditary changes in characters, may have physical bases not at present recognized."[282]

Like Rhoades, Sonneborn recognized the role of these factors in developmental biology. "With the occurrence of such determiners in the cytoplasm and, at least at certain stages, not in the nucleus, all that is required to account for the production of different characters in different cells with the same genes is to have differential segregation of these cytoplasmic determiners at cell division. . . ."[282]

These cytoplasmic factors in maize and Paramecium stimulated considerable interest. C. D. Darlington, in 1944, called them plasmagenes and he endorsed the implications that Sonneborn and Rhoades had listed. "The frontiers that exist between the studies of heredity, development, and infection are . . . technical and arbitrary, and new possibilities of analysis and experiment will arise when we have to learn the passwords to take us across them."[94]

In the mid-1940's Sonneborn pursued the study of the killer factor and other cytoplasmic or non-Mendelian traits in Paramecium. There were, in addition to the killer trait, cytoplasmic phenomena apparently affecting sex determination and antigen formation in one of two strains of P. aurelia explored by Sonneborn and his associates. At first the plasmagene model seemed appropriate for generalizing these new findings as representatives of a widespread phenomenon. But in 1946 J. R. Preer, investigating the kappa factor in Sonneborn's laboratory for his Ph.D. thesis, developed a technique for diluting out the kappa particles from killer strains belonging to a different variety with somewhat different properties. "By taking advantage of the differential rate of increase of the particles of the cytoplasmic factor in relation to the rate of fission of the paramecium, it became possible . . . to free the paramecia from the particles and so produce strains permanently lacking them and the character they control; and . . . to induce cultures which had lost the character because of considerable reduction (but not total loss) of their particle number, to reacquire the normal number of cytoplasmic factor particles and the character associated with them. Finally a technique was devised which demonstrated that a single cytoplasmic factor particle in an animal is sufficient to give rise to the normal number of particles." Additionally, Preer was able to calculate "the number of particles as approximately 256 in one particular individual and also to calculate that the duplication rate of the particles was about 1.9 per day while the paramecium were dividing 3.3 times per day."[264]

Before Preer had discovered the number of particles in a killer, Sonneborn had assumed that these were about the size of genes and that they might have evolved from them. If kappa were a descendant of gene K, this theory could be extended to a model of differentiation and metabolism in which nuclear genes produced counterpart cytoplasmic genes. "The conception of the gene consisting of two parts was proposed as a working hypothesis. One part, localized in the chromosome, was assumed to function primarily in the process of self-duplication; the other part, occurring . . . in the cytoplasm . . . was assumed to be the physiologically active part of the gene, entering into the determinative processes of cellular physiology. Each cytoplasmic factor of Paramecium was assumed to be the free and active part of a gene. . . . It was apparently an error to assume that because the cytoplasmic factors were originally derived from the nuclear genes, they could reunite with these genes under certain circumstances in the macronucleus. This latter part of the hypothesis has nothing to support it any more. However, the comparison between the two groups of varieties of *P. aurelia* still suggests that the origin of cytoplasmic factors in Paramecium is to be traced ultimately to the nuclear genes."[283]

Preer's discovery, however, led to a strong reaction against the plasmagene theory and the possible evolutionary origin of the plasmagenes from the nuclear genes. Edgar Altenburg was one of the first to criticize it. "In view of the recent findings just reported by Dr. Sonneborn, it is no longer necessary to assume any direct relation, in point of origin, between K and kappa. One can account for the experimental findings by assuming that K arose first by mutation from k, and that kappa later arose by mutation, independently of K, from self-reproducing bodies normally present in the

cytoplasm. It might be pointed out here that the number of green symbionts in Paramecium bursaria is about 250, and that this is the same as
the number of the bodies responsible for the killer effect in Paramecium
aurelia. . . . This remarkable correspondence in number suggests the possibility that the bodies in question are related to the green symbionts,
having possibly been derived from green symbionts in a common ciliate
ancestor."[2] Sonneborn, however, did not agree with this symbiont theory
of Altenburg's or a similar viral origin suggested independently by C. C.
Lindegren. In a viral infection there would be an immense number of
particles, but there were far fewer kappa particles per cell. No bodies
comparable to the green chlorellae in P. bursaria, were seen in the P. aurelia
killers. Most significantly, a killer strain of P. bursaria had been found.
"When first received, this killer P. bursaria had typical green algae, eventually it lost its algae but remained nevertheless a killer."[283]

The association of killing with the presence of kappa led to an interesting biochemical interpretation of the function of kappa. "As the volume
of these paramecia is roughly 200,000 cubic microns, a normal killer cell
contains, on the average, about one particle of kappa per 1000 cubic
microns. If, as seems probable, the particles of kappa are single molecules
distributed essentially at random throughout the cytoplasm, each molecule
of kappa in its 1000 cubic microns of cytoplasm would be, for all practical
purposes, isolated from every other kappa molecule. The activity of such
isolated single molecules would be completely outside the scope of the
chemical laws of mass action. Here at least is one substance important in
cellular differentiation that seems to demand for its proper understanding
a new kind of biochemistry, a biochemistry of single molecule activity."[284]

By 1948 several new findings had been made which further weakened
the extension of the plasmagene theory. Sonneborn had found that one
of the new cases of cytoplasmic inheritance, sexuality in a strain of P.
aurelia, was not due to a factor similar to kappa but to a cytoplasmically
located product of the parental macronucleus which acts upon new macronuclei developed from a product of the fertilization nucleus. Preer had also
used radiation in a target theory test of the size of kappa particles. His results
were unexpected. "It is likely that the inactivation dose of kappa . . . indicates
kappa size. . . . It is probable that kappa is at least as large as the largest
viruses (0.1–0.3 μ) or bacteria (0.15–3 μ). Particles of this size should be
visible with the compound microscope."[265] Sonneborn accepted these findings: "These new results . . . lead to the conclusion that kappa cannot be
distinguished from a virus even on the grounds of mode of transmission. . . .
What these agents are to be called seems less important than obtaining an
understanding of their nature and properties such as is provided by the work
of Preer, Dippel and ourselves, and obtaining evidence as to whether they are
normal and integral parts of the genetic system."[285] Studies of antigenic
traits, however, "support the view that the cytoplasmic factors or plasma-

genes of Paramecia are, unlike viruses, normal and integral parts of the genetic system."[285]

If they were "normal and integral" components of cellular heredity, the plasmagene theory was still defensible, even if the particle size suggested a body comparable to a chromosome or a segment of a chromosome. "Plasmagenes differ from nuclear genes, so far as now known, only in their location in the cytoplasm instead of in the nucleus. . . . Eventually, if the plasmagene hypothesis proves satisfactory, the two special theories for nuclear genes and for plasmagenes will have to overlap to the extent of stating the relation between the two kinds of genes and their division of labor. At present, some believe each kind of plasmagene to be an exact copy of the nuclear gene that controls its production; others believe there is only a general, chemical family resemblance between them. Regardless of whether the twin genes, nuclear and cytoplasmic, are 'identical' or 'fraternal,' their separation into different environments has probably had profound consequences. Confinement of nuclear genes to the chromosomes made them the vehicles of Mendelian heredity. The less confined and more variable cytoplasmic environment of plasmagenes perhaps permits them to serve the varied demands of cellular differentiation, the great enigma of orderly development from eggs to adult."[286]

Kappa was stained and made visible for the compound microscope in 1948. Also, Sonneborn's associate, van Wagtendonk, demonstrated the presence of DNA in kappa. These facts excluded the possibility that they were mere replicas of chromosomal genes, but it did not eliminate their significance in interpreting cytoplasmic phenomena. Sonneborn still favored the hypothesis in 1950. "Cytoplasmic inheritance among both plants and animals, and in multicellular as well as unicellular organisms is . . . not an hypothesis but a fact—one of the capital facts of biology. Hypothesis and opinion enter only when this fact is interpreted, when the physical bases of cytoplasmic inheritance and the mechanisms by which they operate are discussed. . . . The existence of fundamental differences between different examples of cytoplasmic inheritance (is) . . . not immediately apparent. Indeed the different examples may at first seem to be similar in principles. I have learned this the hard way. . . . Apparently more than one model is needed to account for the various examples of cytoplasmic inheritance."[289] The unexpectedly large size of kappa was not a decisive argument against plasmagenes. "I shall employ the term plasmagene for those cytoplasmic structures known to manifest genic properties however infelicitous this may seem when applied to bodies large enough to be microscopically visible. . . ."[287] Sonneborn suggested that kappa had become a normal element in the cells which absorbed it. In this way, kappa particles resembled plastids whose inheritance was not dismissed as insignificant on the grounds of a possible symbiotic or infectious origin in their early evolution. Sonneborn also favored the extension of the plasmagene interpretation to the ciliary antigens of Paramecium. "Two considerations favor the hypothesis of gene-specified plasmagenes (supplemented with a competition or inhibition assumption). First, some of the discoveries made in the serotype work were predicted on the basis of plasmagenes and competition . . . second, plasmagenes are known to be the basis of cytoplasmic inheritance in a number of cases . . . while not a single example of cytoplasmic inheritance has yet been shown to be based on variable gene activity or mutual inhibitions or competition and autocatalysis."[287]

By 1955, however, Sonneborn found little support among geneticists for the plasmagene model, "Although at first hailed as the best, almost the only, example of cytoplasmic inheritance in animals, kappa is now looked upon by most geneticists as a foreign organism."[288] There were also other unsuccessful attempts to use the plasmagene theory for cytoplasmic phenomena. The sigma factor in Drosophila killed flies carrying it when they were exposed to certain concentrations of CO_2. The "petite" colony in yeast respired anaerobically but not aerobically, with all oxidative enzymes apparently affected en masse. For sigma, a virus was inferred; for the "petites," either the loss or the mutation of mitochondria was proposed.

In 1959 Sonneborn reviewed the work on kappa. The particle was not a virus; it was sensitive to antibiotics. It was not a typical alga since it lacked a nucleus and did not contain light-sensitive pigments. It lacked the chromatin bodies found in typical rickettsiae and bacteria, although a faint condensation similar to chromatin bodies was present in electron micrographs. The presence of enzymes in kappa was disputable, which would favor the viral interpretation. The stains for bacterial spores were ineffective on kappa. "It thus looks as if either kappa is intermediate between viruses and bacteria or some of the conclusions concerning its organization and biochemical properties will require revision after further study. . . . The initial problem raised by kappa . . . appears to have been solved. They are cytoplasmic genetic factors of Paramecia by virtue of being hereditary symbiotes and are as well integrated into the genetic system of paramecium as bacteriophage is in the genetic system of bacteria."[289] While kappa was thus abandoned as a model system for cytoplasmic inheritance, it retained its value for the study of "hereditary endosymbiosis." In 1961 this model of kappa for host-parasite relations was reaffirmed. "The particles . . . are certainly not viruses; they are infectious, intracellular parasites manifesting a more complex level of organization which is close to, if not identical with, that of rickettsiae and bacteria."[290]

The significance of the kappa studies as well as the plasmagene theory was its effect on genetic thinking, rather than its essential validity. While kappa were not plasmagenes as they were inferred to be during the early 1940's, they directed attention to the problem Morgan hoped to revive. Microorganisms were demonstrated to be important for exploring cellular differentiation. The investigation of cytoplasmic traits, initially stimulated by the plasmagene model, led to new discoveries of chromosomal behavior such as macronuclear reorganization in *Paramecium*. The antigen system led to theories of alternative steady states, of gene action which could be switched on and off by cytoplasmic systems. Novel cytoplasmic phenomena such as the inheritance of cellular structure, were discovered because of Sonneborn's attention to the cytoplasm as a relatively unexplored area. Similarly, the field of nucleocytoplasmic relations was stimulated by the experiments using the special mating features of *Paramecium*. The implantation of nuclei into enucleated cells, particularly in higher organisms, is in large measure attributable to the plasmagene hypothesis. If the plasmagenes did not turn out to be "the partners of the genes," they at least led to a view of the cytoplasm as more than a "playground for the genes."

CHAPTER 21

Pseudoallelism Versus Intragenic Recombination

Although Sturtevant and Morgan in 1913 had conceptually looked upon the mutants *white*, *red*, and *eosin* as a possible case of closely linked genes, their choice of recombination tests was based on a pair of alleles which were destined *not* to recombine—white eyes, the presumptive double mutant, and red eyes, the presumptive double normal. The discovery of other members of this white-eye series of multiple alleles did raise this question again. In 1913, S. R. Safir discovered the mutant *cherry*, which was tested for recombination with white and with eosin. The sample size in each case was small and no recombinants were found. This was also true for Safir's tests of another allele, *buff*, found in 1916, which he tested with eosin, white, and cherry. Another student of Morgan's, R. R. Hyde, reported two new alleles, *tinged* and *blood*. Ten different combinations of these alleles were tested, using fairly sizable numbers in some series (18,000 in the white × blood cross), but without success. "The assumption that these . . . mutants are the result of changes in loci lying very closely together on the chromosome as demanded by the Presence and Absence theory has been tested by Morgan and others by means of their linkage relations in three possible combinations. . . . The discovery of two new mutants has made it possible to carry out the test in eight additional ways. The evidence, which involves data from something like a half million animals weighs heavily against the Presence and Absence theory and is entirely in accord with the assumption that something analogous to isomerism may change an hereditary factor resulting in the production of a new form."[166] Actually, the information is misleading because at least half the

184

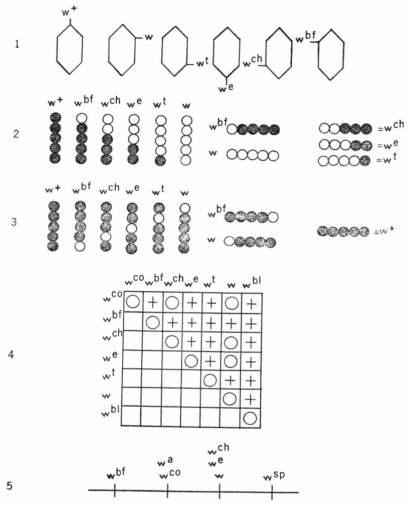

Figure 27. Pre-pseudoallelic Models of the White Locus. Morgan's students, Hyde and Safir, attempted to obtain crossovers from pairs of white alleles. In model 1, none would be expected. In 2, a quantitative series of alleles would parallel a quantitative increase of affected sub-units. Recombination between buff and white would provide a series of allelic types. In 3, the site of the sub-unit was assumed essential for phenotype. This would lead to wild-type recombinations from buff and white compounds. In 4, the various alleles tested by Hyde and Safir are indicated by a +. The map of the white locus is shown in 5. Note that many of the tests made by Hyde and Safir could have produced wild-type crossovers.

progeny were derived from crosses of red with different alleles of the white series. It is also a curious accident of fate that those series which would have yielded crossovers were among those in which Hyde and Safir used the smallest numbers of progeny. Conversely, the greatest expenditure of time was on alleles which were not separable because they occupy the same site or are so close together that their resolution has not yet been achieved. Examination of all the crosses in the white series by Safir, Hyde, Sturtevant, and Morgan suggests that the probability of having obtained at least one crossover was slightly in their favor. This negative evidence discouraged any further attempt by Morgan's group to pursue such tests. "When white, tinged, eosin, cherry, and blood are crossed with each other, no red eyed flies appear in either the F_1 or F_2 generation. This fact when taken in connection with the independent origin of the five factors from red, proves that they form a system of multiple allelomorphs."[167]

That the white-eye series might represent an integrated segment of the chromosome was proposed by East in 1929. "There have been twelve mutations at the so-called 'white locus' in Drosophila melanogaster. Let us assume that a change has taken place in each case at a different 'spot' in a linear series of twelve 'spots' all so close together, or for some reason bound so strongly together, that crossing-over never takes place. A change takes place at No. 1 link which produces white, a change takes place at No. 2 link which produces pink, a change takes place at No. 3 link which produces eosin. Now Morgan says that 'if eosin arose in this way, then, when such an eosin fly is crossed to the original white stock it should give red, since the effect of the new recessive eosin in one chromosome of the F_1 is cancelled by the effect of the normal allelomorph of eosin.' This is either imagining too much or not enough. The change in the No. 1 link gave white without respect to what remained. The change in the No. 3 link gave eosin without respect to what remained. A cross between the two would give whatever the combined effects of the haploid white and haploid eosin happened to be; and only those forms would be recovered in the F_2 generation, since no crossing over has been postulated. . . . Such a question holds little interest for theoretical genetics at present, but it does bear on the nature of the gene."[121] East, however, fell back on his earlier view that the nature of the genes was still unknown and they could best be described as concise units without physical reality. "In terms of geometry, chemistry, physics, or mechanics we can give them no description whatever."[121]

Other events eclipsed interest in this problem. Large numbers of multiple allelic series were discovered in the twenty-five years after the Drosophila group abandoned its search for crossovers between alleles. In none of these cases was a deliberate search made for crossing over. In 1940, however, C. P. Oliver reported "a reversion to wild-type associated with crossing over in Drosophila melanogaster." Oliver used the lozenge-eye mutants, glossy and spectacle, in a test of fertility of the compounds of lozenge alleles. Among the progeny from such partially fertile compounds were eleven cases of "reversion" to the wild-type. "In a study of two alleles of the lozenge eye in *Drosophila melanogaster*, a low frequency of reversion to the wild-type has been discovered. The reversion involves the color and structure of the eye, and also the genital tracts which are abnormal in the mutant females. In each case in which the reversion has occurred, crossing over between the X-chromosomes has also occurred."[256]

The similarity to the Bar case was only partial. "Although crossing over seems to be associated with the reversion, and one crossover type appears, the complementary type has not been recovered. It is not known whether the failure to recover the complementary type is due to the inability of that type to live, or to the inability of the observer to recognize the combination. . . . Although crossing over is an active factor in the reversion of the alleles to the wild type, it is not possible as yet to determine the exact nature of the phenomenon. The condition can be a case of unequal crossing over; but it can as likely be a case which involves the 'repeat' hypothesis developed by Bridges, in which different primary loci of the chromosome are involved in the expression of the two mutants."[256]

This repeat hypothesis, based on the analysis of salivary chromosome structure, proposed an evolution of genes by tandem duplications (presumably by the same "primary unequal crossing over" advocated by Muller) with their subsequent independent genetic evolution. E. B. Lewis the next year reported another case, involving the dominant mutant Star and its recessive allele (later called asteroid). "The situation may be similar to a case, reported by Oliver, of reversions, associated with crossing over in one direction, arising from females carrying two alleles of the lozenge eye mutation. For the sake of simplicity, S and S^r are considered as alleles. . . ."[180] The following year Lewis reported the isolation of the double mutant, and the recessive Star was renamed asteroid, ast. In this analysis he reported sixteen cases of the double normal $+ +$ and three cases of the double mutant S ast. This meant that there were two complementary classes of crossovers (in one direction, *not* unequal crossing over) and the phenotypes of the two arrangements S ast $/ + +$ and $S + / +$ ast were different. The $S + / +$ ast arrangement showed a more abnormal eye than did the S ast $/ + +$ arrangement.

In 1945 Lewis made use of Bridges's and Muller's theories of gene evolution for an interpretation. "A plausible interpretation of the S and ast loci can be developed by assuming that they have resulted from duplication of an ancestral locus, that duplication now being established in the species. This notion is chiefly based on the finding that S and, very probably, ast are included in the 21E1–2 doublet structure of the salivary gland chromosomes. . . . If doublet structures are repeats, as the evidence thus far indicates, then, judging from their widespread occurrence in the salivary gland chromosomes of Drosophila, it is likely that other multiple

allelic series may be resolved into duplicate loci which act, by reason of
a position effect, as a developmental unit."[182] This secondary assumption
of a position effect was invoked by Lewis to account for the differences in
the two arrangements of the alleles. The cis arrangement a b / + + was
more normal than the trans arrangement a + / + b, which gave a strong
mutant expression. In 1948 Lewis reported two additional cases, involving
the Stubble bristle and the bithorax abdominal series. In each of these two
cases the cis-trans differences were obtained. As in the lozenge and Star
cases, the various alleles were distinguishable from one another in their
positions on the map as well as in their character expression. These findings strengthened Lewis's interpretation that they had arisen by duplication and were now in the process of evolutionary change, their similarity
of function maintained by position effect.

The phenomenon found by Lewis was called pseudoallelism in 1945
by McClintock, who reported a superficially similar situation in maize.*
As Lewis explored this phenomenon in the several cases he had detected,
he saw its direct application to gene evolution. The presence of a cis-trans
difference in appearance defined the existence of such related genes which
he called "position pseudoalleles." At the same time Beadle's work on
biochemical pathways stimulated Lewis to propose a model of sequential
steps in the synthesis of a final product, but in this hypothetical pathway
the intermediate products were immediately used by the neighboring pseudoalleles. In the trans case each chromosome would thus be blocked in the
biochemical pathway at either one of two possible points. The lack of a
diffusible substance prevented one intermediate from reaching the homologous chromosome. In the cis arrangement one chromosome, with the two
normal alleles, could complete the pathway and thus the cis arrangement
would tend to be more normal than the trans arrangement. "The possibility that duplicate genes may often diverge from one another in their
functioning in the above way is an attractive one since it gives a conservative, and yet, progressive process such as is required for a general theory of
gene evolution. Indeed . . . the development of sequential steps at this level
may only be possible in the case of genes which were once identical."[184]

In Great Britain a different view of gene structure and integration
was being developed by G. Pontecorvo and his students. Pontecorvo had
emigrated from Italy to Scotland, where he studied with Muller, who had
emigrated via Spain from the U.S.S.R. This was the period of Muller's
intense interest in the problems of gene structure and the left-right test.
Pontecorvo, after the end of the war, directed his attention to this problem,
using his own interests in biochemistry to formulate a theory of pseudoallelism. In Pontecorvo's theory, the repeat origin of pseudoalleles was not

---

* McClintock studied the alleles yellow, pale green, and white. Yellow and pale
green were complementary in the compound, but each manifested allelism with white.
White and pale green were cytologically demonstrable deletions, but yellow was a point
mutation.

A. THE LEWIS PSEUDOALLELIC MODEL

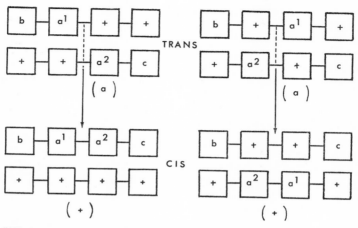

B. THE PONTECORVO INTRAGENIC MODEL

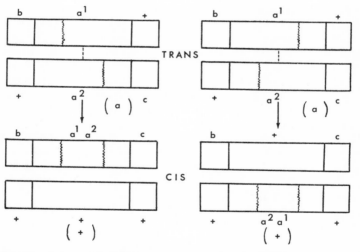

Figure 28. Intra- and Interallelic Models for the Cis-Trans Effect. In A, the genes (pseudoalleles) a^1 and a^2 are mutant in trans because of a position effect. The intact sequence in cis restores normal function. In B, the genes a^1 and a^2 are unitary but the lesions in their mutations are at different sites. Recombination *within* the gene results in one intact sequence with a cis form (a^+) and a double mutant. In both cases the position of a^1 and a^2 may be determined by the placement of outside markers (b and c). The two possible arrangements of a^1 and a^2 are indicated. This "four point test" is used for ordering the sites (or pseudoalleles) in a multiple allelic series.

necessary. Any closely linked genes could become sequentially related if the intermediate products were used up immediately, before they had a chance to diffuse away from the chromosome. Thus, Pontecorvo, like Lewis, invoked a model of position effect similar to the "immediate localization of gene products," as Muller had phrased it. The position effect also could be related to the model of biochemical pathways used and extended by Beadle's investigations in Neurospora. Pontecorvo described such apparent non-diffusibility as "millimicromolar reactions," a concept earlier developed by McIlwain. If selection established a set of millimicromolar reactions along the length of the chromosome then there would be no need to limit the universality of pseudoallelism. Pontecorvo's model implied a more widespread occurrence of pseudoallelism than Lewis's model, whose upper limit of pseudoalleles was determined by the maximum number of salivary chromosome doublets.

With the aid of Pontecorvo's hypothesis, J. A. Roper initiated "a search for linkage between genes determining vitamin requirements" in the mold, Aspergillus nidulans. "A working hypothesis for the deliberate search for linkage is that close linkage might be expected between some of the genes acting on any one series of millimicromolar biochemical reactions. A first deliberate search for linkage prompted by this working hypothesis has led to positive results. . . . Each of three mutant strains of Aspergillus nidulans requires biotin or desthiobiotin for growth but does not respond to pimellic acid, and attempts at further characterization of their growth-factor requirements have so far been unsuccessful. . . . The three alleles are not allelomorphic of one another, their loci are in a chromosome segment about 0.2 Morgan in 'length' and the order of these loci has been established. . . . That a first deliberate search based on the above mentioned working hypothesis should be successful is scarcely likely to be a matter of chance."[270]

Pontecorvo reconsidered the mechanism for pseudoallelism in 1952. The success in finding recombination between multiple allelic series in Aspergillus raised the possibility that this was a universal characteristic of all alleles. "The most reliable method of resolving the chromosome into its linear array of genes is that of crossing over. So far, the resolving power of crossing over is limited only by the maximum number of products of meiosis which can be analyzed."[259] If this were true, then the absence of crossing over in a given pair of alleles would only indicate a lack of sufficient effort by the investigator himself! By comparing the minimal crossover distances in Aspergillus with the total length of the map and the corresponding chromosome volume, Pontecorvo recalculated the estimates of the gene size that Muller and Morgan had attempted. "The estimates of the size of a gene by means of genetic techniques give maximum lengths of the order of thousands of angstroms. The target volume estimates give minimum diameters of tens of angstroms. The difference seems to be too great to infer that the two may eventually coincide. . . . A tentative conclusion could be that the gene as a unit of physiological activity is based on a chromosome segment longer, in certain cases at least, than any one part of the chromosome where mutation may occur. This means that either a gene has only one mutable site or unit, constituting a small portion of it . . . or it has several mutable sites, changes at any one of which all produce the same change or similar changes, in physiological action."[259]

It is interesting that Pontecorvo retained an interpretation similar to Muller's in the discussion of the left-right test: the possibility of functional discontinuity as an alternative to the precision of breakage boundaries in the scute region. But, unlike Muller, Pontecorvo accepted the general findings of the target theory instead of discarding them on the basis of the inherent assumptions used in the model. From the association of a large number of "target" sites within the gene, and the ease of obtaining recombination between different alleles in a multiple allelic series, Pontecorvo suggested an alternative interpretation of gene structure. "Crossing over may occur within this length of a chromosome between the various mutable sites: thus the ultimate unit of crossing over and the ultimate unit of mutation—which do not need to be the same thing—could be at least in some cases one or two orders of magnitude shorter than the chromosome segment which forms the basis for the unit of physiological action."[259] This view, advocating intragenic recombination and the continuity of the genetic material for physiological activity, was similar to the view of the *entire chromosome* proposed by Goldschmidt. More significant, however, was its implication for the gene concept. "We are dealing with one gene and the . . . alleles are due to mutation at . . . different mutation sites of that gene, in every case the result of mutation being that of inactivating the gene. On this interpretation, in a heterozygote for two *different* mutant alleles, recombination between mutational sites could, and does in fact, occur. . . ."[259]

The direct application of this theory to *Drosophila* was made by MacKendrick and Pontecorvo. They chose the long neglected *white* locus. "The list of reported cases of crossing over between alleles in other organisms is growing very rapidly, and a pressing question is how general a property of genes this happens to be. In some of the examples available in the literature, the loci investigated were studied thoroughly precisely because in the course of previous breeding experiments unexpected types had been obtained. . . . These *selected* loci cannot be included, of course, in a survey to infer how widespread crossing over between alleles is. . . . We thought an ideal *unselected* locus for a search for crossing over between alleles was *w* in Drosophila: at this classical locus a large number of alleles are known, with distinguishable and pleiotropic effects, suitable closely linked markers can be placed on both sides and the cytological situation does not suggest any peculiarity."[190] Their success in obtaining crossovers among alleles of the white locus supported "the idea that this is a very widespread property of genes."[190]

That same year, 1952, Lewis studied the white locus independently of Pontecorvo. His reason for choosing this region was his belief that a fairly large number, but not all of the multiple allelic series would prove to be pseudoallelic. "With the aid of more adequate techniques for studying crossing over than were available in the early studies, the white gene and its so-called 'allele,' apricot, have been reinvestigated. . . . These two genes occupy separate loci and . . . they constitute another example of position pseudoallelism. . . . The apricot gene, formerly symbolized as w^a will be designated by a new symbol, namely, apr."[185] Lewis's frequency of wild-type recombinants was 12/40,000 which was about six times higher than would be obtained without the special inversions used to boost the cross-over frequency. This meant that white and apricot would have a frequency of about 1/20,000 without the use of inversions in other chromosomes. Thus, in several of the series used by Hyde, the probability of obtaining a crossover was as good as that of not obtaining one. If by chance, apricot or cherry had been found before eosin, the tests run by Morgan, Sturtevant, and Bridges would have yielded about seven reversions to wild-type from the allelic compounds they tested.

The interpretation offered by Lewis, however, was still unmodified from his earlier views. "This result suggests that a mutant gene at one of the loci blocks or impairs the functioning of the normal allele of the gene at the other locus, when both are present in the same chromosome, as in the apr + / + w heterozygote; while no impairment of the functioning of the two different wild type alleles is phenotypically detectable when both of these are present in the same chromosome. The simplest assumption is that the effect is one way; that is, that the mutant gene apr impairs the functioning of w^+, or that w impairs that of apr^+. This leads to a simple model in which one of the genes controls a step $A \rightarrow B$, and the other a step $B \rightarrow C$, in a biochemical reaction chain of the type $A \rightarrow B \rightarrow C$. The position effect can then be assumed to result from a failure of substance B to diffuse readily from one chromosome to the other so that the chain of reactions in one of the chromosomes of the heterozygote is carried out more or less independently of the chain in the homologous chromosome."[185] As he had proposed earlier for Star, bithorax, and Stubble, Lewis favored the doublet theory of the origin of the white pseudoalleles. "Cytological studies of Panshin and others have shown that the w^+ gene is located within the confines of the 3C2–3 doublet, or two banded structure of the salivary chromosomes. . . . The probable significance of this cytological finding is that the pseudoallelic genes associated with a doublet represent an established duplication of a single ancestral gene."[185]

Pontecorvo rejected this viewpoint. "In earlier genetics the tacit assumption has been that the three ways in which the gene could be defined

—unit of crossing over, unit of mutation and unit of physiological action—
were co-extensive. Even the discovery by Sturtevant . . . of the position
effect did not lead to the rejection of this tacit assumption; the position
effect could be considered the consequence of localized interactions between distinct genes. . . . We owe it to Goldschmidt . . . to have stressed
for long that, on the contrary, the position effect made it necessary to reappraise our views on gene structure and action. But we owe to Muller
. . . the clearest statement of the previous conceptual limitations."[260] The
cis-trans phenomenon now seemed universal to Pontecorvo. "I propose
that this phenomenon should be called the 'Lewis effect' instead of 'position pseudoallelism,' as E. B. Lewis himself suggested because he was im-
plying a special rather than a ubiquitous feature of alleles."[260]

In virtually every case studied by Pontecorvo in *Aspergillus*, the gene
responded physiologically in the same way to a mutational change no matter where the damage had occurred. One case, however, did provide something unexpected. Instead of a trans compound expressing the mutant
condition, two closely linked sites affecting the same biochemical step were
complementary in trans. Such "complex loci," similar to the dumpy and
Notch cases reported in the 1920's, required a modification of the universal
principle of allelism associated with the "Lewis effect." "Clearly, the distinction between alleles within one of the regions mentioned before, and
alleles at 'complex loci' . . . is one which rests exclusively on mode of
action. In one case we have the Lewis effect, in the other case we have
complementarity. . . . For a working model of the organization of the chromosome over minute regions it may be useful to assume that the difference
between the two types of relation is a consequence of spatial organization,
in first approximation, of distance apart. In a very crude way, it could be
supposed that mutational sites very close to one another give origin to
alleles showing the Lewis effect, and mutational sites further apart give
origin to alleles showing complementarity. The assumptions behind this
model are the ones I proposed some years ago. . . . If we consider stepwise
reactions occurring on the surface of a chromosome in an assembly line
fashion—i.e. ordered in space sequence correspondingly to the time sequence—a rate of millimicromolar order of reactions, instability, or non-
diffusibility of the intermediates could all account singly or in combination
for the Lewis effect. However, the Lewis effect could no longer operate
when the distance apart between the two mutational sites is greater than
the average distance between two homologous chromosomes."[260] The
return of the millimicromolar system in this interpretation was not a
replacement, but an extension of the intragenic model of recombination
for allelic series. It was required for the interpretation of the complementation existing among some of the alleles of a multiple allelic series.
The model explicitly predicted a genetic test: a proportionality should
exist between the distance of alleles and their degree of complementation.

Lewis, of course, was aware of the "complex loci" in *Drosophila*. They existed in one of his pseudoallelic series—the bithorax series. In 1954 he offered a model to account for the difference between position pseudo-allelism and complementation. "The position effect in position pseudo-allelism has been interpreted by assuming that the genes concerned control successive steps in a sequential reaction series of the type: A → B → C; and that the substance B, at least, diffuses appreciably only along the chromosome. . . . It follows on such a model that somatic pairing might act as a modifier of this type of position effect and would tend, by bringing homologous chromosomes close together, to minimize the difference between the $a\ b\ /\ +\ +$ and $a\ +\ /\ +\ b$ heterozygotes. These considerations have led to a study of the effect of chromosomal rearrangements on certain pairs of mutant genes belonging to the bithorax series. . . . These results show that, as predicted on the above model, the phenotype of the $a\ +\ /\ +\ b$ type of heterozygote can be quite significantly altered in the direction of a still more extreme departure from wild type by the presence in heterozygous condition of a broad class of chromosomal re-arrangements; namely, the majority of those having at least one breakage point in the region from the centromere of the third chromosome to the locus of bx (89E1–4)—a region comprising over 500 bands of the salivary gland chromosomes. On the other hand such rearrangements do not exert a detectable phenotypic change on the $a\ b\ /\ +\ +$ heterozygote, nor on either the $a\ /\ +$ or $b\ /\ +$ heterozygotes, indicating that their effect is not exerted directly on the genes themselves."[187]

Lewis rejected Pontecorvo's model as inadequate for explaining this new phenomenon in *Drosophila*. "These . . . studies are at variance with the speculation that the whole region occupied by a pseudoallelic series acts as a functional unit; instead they demand a particulate interpretation in which genes at the individual loci of the pseudoallelic series are distinguishable from one another not only by crossing over but by function, as well." This remarkable pairing disturbance Lewis called a "transvection effect." Position pseudoallelism, in contrast, involved a different position effect. "The position effect which is detected by comparing the cis and trans type and which is the basis for defining position pseudoallelism now needs a term to distinguish it from the transvection effect, and will henceforth be referred to as the 'cis-vection effect.' "[187]

Thus, in 1955 two models were available for analysis of gene structure. One hypothesis assumed the universality of intragenic recombination and the existence of millimicromolar biochemical reactions. The other model assumed the occasional chromosomal duplication of two or more genes whose independent mutations led to their differentiation as biochemically related pathways; diffusibility of gene products between homologues, not the distance along the length of the chromosome, was believed to determine the degree of complementation between pseudoalleles. Ponte-corvo's model correctly predicted, for many organisms, the likelihood of recombination between alleles (although the basis for that was a discredited target theory). Nevertheless, the hypothesis proved to be false for the ad hoc millimicromolar interpretation of complementation. Lewis's model, based on Drosophila work, predicted discontinuity in crossing over; this may have been true for Drosophila but it was not true for other systems. The integrated sequential biochemical pathway which Lewis advocated was regarded with disfavor for at least five years. The universality, or apparent universality, of intragenic recombination, on the other hand, was almost entirely accepted by geneticists during these five years because of the spectacular discovery of genetic fine structure in viruses and bacteria.

CHAPTER 22

Genetic Fine Structure

The scientific contributions leading to "molecular biology" came from many sources, but one stands out as an unexpected and pioneering achievement. This is the biochemical analysis of bacterial transformation. In 1928 Griffith, in England, reported that dead, boiled preparations of encapsulated strains of *Pneumococcus* transformed nonencapsulated pneumococci into the encapsulated type characteristic of the donor. The capsule was a polysaccharide and it was a necessary component for the virulence of this strain of bacteria. Although little was known about the nature of the transforming substance or "principle" in the boiled preparation, the event itself was hailed as a possible example of directed mutation, because the transformed cells retained a permanent change in type. In 1944 a team of microbiologists, Avery, MacLeod, and McCarty, at the Rockefeller Institute, attempted a biochemical analysis of the transforming principle. Their publications presented convincing evidence that they had determined "the chemical nature of the substance inducing specific transformation of pneumococcal types. A desoxyribonucleic acid fraction has been isolated from Type III pneumococci which is capable of transforming unencapsulated R variants derived from Pneumococcus Type II into fully encapsulated Type III cells. . . . The induced changes are not temporary modifications but are permanent alterations which persist provided the cultural conditions are favorable for the maintenance of capsule formation. . . . From the point of view of the phenomenon in general, therefore, it is of special interest that in the example studied, highly purified and protein free material consisting largely, if not exclusively, of desoxyribonucleic acid is capable of stimulating unencapsulated R variants . . . to produce a capsular polysaccharide identical in type specificity with that of cells

196

from which the inducing substance was isolated. Equally striking is the fact that the substance evoking the reaction and the capsular substance produced in response to it are chemically distinct, each belonging to a wholly different class of chemical compounds."[5]

Throughout the preceding four decades of genetic studies it had been assumed, for the most part, that the genic material itself was a protein. This was reasonable because there are many amino acids; each protein has its own immunological specificity, the proteins assume a variety of shapes through folding; certain viruses such as TMV are chiefly proteinaceous; and proteins have numerous metabolic activities. Nucleic acids, on the other hand, are composed of very few elements; isolation and purification of the nucleic acids indicate a uniform, simple, crystalline structure; and for years no known function could be attributed to them. In most biochemical and genetic interpretations, nucleic acids were believed to form a scaffolding on which the more biologically significant proteins of the chromosome were aligned. "If, however, the biologically active substance isolated in highly purified form as the sodium salt of desoxyribonucleic acid actually proves to be the transforming principle, as the available evidence strongly suggests, then nucleic acids of this type must be regarded not merely as structurally important but as functionally active in determining the biochemical activities and specific characteristics of pneumococcal cells."[5]

The following year, 1945, in his Pilgrim Trust Lecture on the gene, Muller suggested a genetic interpretation of transformation. "In my opinion, the most probable interpretation of these . . . Pneumococcal results then becomes that of actual entrance of the foreign genetic material already there, by a process essentially of the type of crossing over, though on a more minute scale. . . . Mirsky gave reasons for inferring that in the Pneumococcus case . . . there were, in effect, still viable bacterial 'chromosomes' or parts of chromosomes floating free in the medium used. These might, in my opinion, have penetrated the capsuleless bacteria and in part at least taken root there, perhaps after having undergone a kind of crossing over with the chromosomes of the host. In view of the transfer of only a part of the genetic material at a time . . . a method appears to be provided whereby the genetic constitution of these forms can be analyzed, much as in the crossbreeding tests on higher organisms. However, unlike what has so far been possible in higher organisms, viable chromosome threads could also be obtained from these lower forms for an *in vitro* observation, chemical analysis, and determination of the genetic effects of treatment."[243]

Another direction of research, initiated by Max Delbrück, made use of bacteriophage as a model system for molecular genetics. The identification of DNA as the genetic material of these viruses was made by Hershey and Chase in 1952. The viruses were labeled with radioactive phosphorus (which is present in the viral DNA but not in the viral protein) and radioactive sulfur (which is present in the viral protein but not in the viral

P³² LABELED BACTERIOPHAGE   S³⁵ LABELED BACTERIOPHAGE

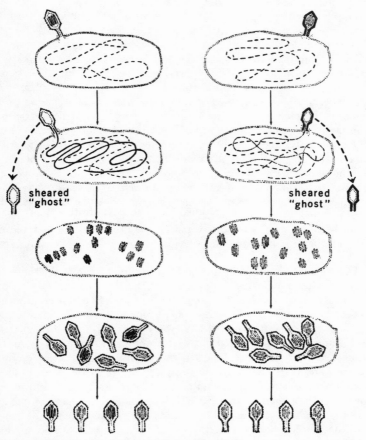

Figure 29. The Hershey-Chase Experiment. DNA contains phosphorus but not sulfur; proteins contain sulfur but not phosphorus. When phage bearing labeled phosphorus infects unlabeled host cells, the labeled DNA is incorporated into some of the progeny. This is not true for labeled sulfur, which remains outside the host as a "ghost." The "ghost" may be sheared off and removed without affecting the synthesis of complete (unlabeled) progeny. Viral DNA, not viral protein, directs the synthesis of new viral protein.

DNA). Such labeled viruses, when infecting unlabeled bacterial cells, introduce radioactive phosphorus to their progeny but not radioactive sulfur. This was interpreted by Hershey and Chase as evidence for a hereditary role of DNA in generating a life cycle; by contrast, no significant role in this process could be attributed to viral protein which served mainly as a protective coat for the virus and which permitted the attachment of the virus to its host cell. Their work convinced most geneticists that the transforming principle was not an isolated example of the genetic activity of DNA; DNA replaced protein as the chemical basis of heredity. The bacteriophage DNA, not its protein, initiated the entire life cycle of the virus subsequent to infection. An intense interest in nucleic acids was stimulated by these new contributions. This was accompanied, in the late 1940's and early 1950's, by a rapid growth of microbial genetics; the extension of genetic principles to viruses and bacteria by Delbrück, Luria, Tatum, and Lederberg required special techniques. The success of these novel techniques led to an enormous increase in "resolving power" for mutation studies, biochemical genetics, and chromosome mapping.

In 1953 the most important discovery in molecular biology was made by J. D. Watson and F. H. C. Crick. "We wish to suggest a structure for the salt of deoxyribonucleic acid (D. N. A.). This structure has novel features which are of considerable biological interest. . . . This structure has two helical chains each coiled around the same axis. . . . Both chains follow right handed helices, but . . . the sequences of the atoms in the two chains run in opposite directions. . . . The novel feature of the structure is the manner in which the two chains are held together by the purine

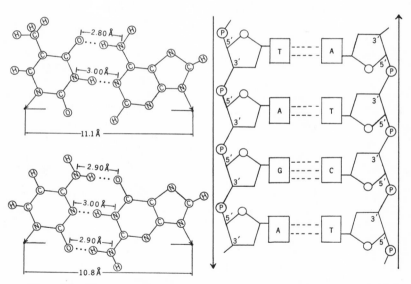

Figure 30. The Watson-Crick Model of DNA. On the left, the base pairing of thymine with adenine (top) is accomplished with two hydrogen bonds. The cytosine-guanine pair (bottom) requires three hydrogen bonds. On the right the anti-parallel direction of the double helix is shown. Each nucleotide is linked by a 3′ to 5′ linkage of the deoxyribose to phosphate. New synthesis would attach nucleoside triphosphates to the 3′ carbon on the lower left or the 3′ carbon on the upper right.

and pyrimidine bases. The planes of the bases are perpendicular to the fibre axis. They are joined together in pairs, a single base from one chain being hydrogen bonded to a single base from the other chain. . . . One of the pair must be a purine and the other a pyrimidine for bonding to occur. . . . If an adenine forms one member of a pair, on either chain, then . . . the other member must be thymine; similarly for guanine and cytosine. The sequence of bases on a single chain does not appear to be restricted in any way."[315] This terse description of the organization of their model was not intended for biochemists alone: "It has not escaped our notice that the specific pairing we have postulated immediately suggests a possible copying mechanism for the genetic material."[315]

In a second short article a few months later, Watson and Crick listed several other implications of their model for genetic theory and testing. "The phosphate-sugar backbone of our model is completely regular but any sequence of the pairs of bases can fit into the structure. It follows that in a long molecule many different permutations are possible, and it therefore seems likely that the precise sequence of the bases is the code which carries the genetical information. . . . Our model suggests possible explanations for a number of other phenomena. For example, spontaneous mutations may be due to a base occasionally occurring in one of its less likely tautomeric forms. Again, the pairing between homologous chromosomes at meiosis may depend on pairing between specific bases."[316] Thus in 1953 several new concepts, partially independent in origin, began to take form. They reflected the attitudes of the different disciplines in which this new generation of geneticists was trained. The excitement of the early years of the Drosophila group was matched by the eagerness of the participants in the new genetics. They were more numerous and more dispersed than their counterparts had been some fifty years before, but they recognized that the fundamental nature of the genetic material might be achieved by use of their new approaches and new techniques. The ideas of an earlier generation of geneticists, limited to genetic and cytological tools, now reappeared in the minds of a new generation, many unfamiliar with the older literature but well aware that a new, molecular, genetics was coming into being.

One of the first successful reports from this school of molecular genetics described the structure of genes in bacteria. The experiments came not from the laboratory of one of the newer generation, but from the laboratory of M. Demerec. Demerec had discarded Drosophila as a suitable organism for genetic studies in the mid-1940's when prolonged attempts to obtain meaningful mutation rates, both spontaneous and induced, were frustrated by the time and expense required for large-scale experiments. In bacterial studies, however, Demerec readily obtained mutations which he could characterize biochemically. The organism used for most of Demerec's first mutation studies with bacteria was E. coli, but suitable genetic techniques were lacking for an analysis of those mutants most likely to be multiple alleles. Fortunately, in 1952, Zinder and Lederberg provided the decisive technique for the analysis Demerec had been waiting to make. "Until last fall (1953) we had used strain B of *Escherichia coli* exclusively in our studies of spontaneous and induced mutability. A serious drawback was that methods were not available in this material for making standard genetic analysis, such as determinations of whether or not similar mutants are allelic. . . . We started work with *Salmonella typhimurium* because Zinder and Lederberg had shown that transduction can be used to determine allelic relationships, and because from the technical standpoint Salmonella is very suitable for our studies."[109]

In bacterial genetics the biochemical mutants that lack a capacity for growth on minimal medium, but which are otherwise viable on supplemented medium, are called auxotrophs. In a transduction test, certain strains of viruses can infect normal bacterial cells and carry fragments of the host chromosomes. These fragments are similar to transforming fragments of DNA. When recipient auxotrophs are infected, the cells are immune to the effects of the virus but they incorporate the fragment brought in. If the fragment happens to carry the normal allele of the mutant auxotroph, such a cell becomes transduced in a manner analogous, if not homologous, to transformation. Thus "similar auxotrophs can be separated by transduction tests into well defined groups. Between members of the same group, transduction either does not take place or is significantly less frequent than between these members and auxotrophs belonging to another group, or wild-type bacteria. The results also show that the grouping based on tranduction tests coincides with that obtained by biochemical methods which investigate blocks in the chain of reactions leading to the synthesis of the compounds required by the auxotrophs. These results favor the assumption that members of each such group are allelic to one another and that the occurrence of a small amount of transduction within a group can be explained on much the same basis as the infrequent recombination that takes place between pseudoalleles. They suggest that a gene locus extends over a section of a chromosome, and that changes occurring in different regions of this section give rise to different alleles. They also indicate that regions within a section may separate, and recombine—by a process analogous to crossing over—with homologous regions within a locus of another chromosome."[109]

Independent of Demerec's first experiments, reporting recombination in bacteria, was an even more elaborate experiment with bacteriophage reported by Seymour Benzer in 1955. Benzer, a physicist, turned his attention to genetic problems after reading a fellow physicist's discussion of the gene concept. The possible application of target theory to genetics and the biophysical concepts which Delbrück and Timofeef-Ressovsky had published stimulated Erwin Shrödinger to write a highly popular book, *What is Life?* Shrödinger's book, in turn, changed the direction of Benzer's interests. After learning the basic techniques for conducting bacteriophage experiments at Cold Spring Harbor, Benzer began an intense study of the problem of allelism in this system.

Benzer chose a phenotype based on morphology rather than biochemistry. Certain mutations produced a larger plaque or colony on a Petri dish appropriately covered with E. coli. In ordinary viral disintegration or lysis of these host cells, the plaque is smaller because of an inhibition of lysis by the large numbers of viral particles outside the bacterial cells. This "lysis inhibition" does not exist in the case of the larger, rapid lysis (r) plaques. As in the case of eye color in Drosophila, several regions of the chromosome are involved in lysis inhibition. Benzer chose the rII region of bacteriophage T4. "A high degree of resolution requires the examination of very many progeny. This can be achieved if there is available a selective feature for the detection of small proportions of recombinants. . . . Such a feature is offered by the case of the rII mutants of T4 bacteriophage. . . . The wild type phage produces plaques on either of two bacterial hosts, B or K, while a mutant of the rII group produces plaques only on B. Therefore, if a cross is made between two different rII mutants, any wild type recombinants which arise, even in proportions as low as 10^{-8}, can be detected by plating on K."[38] This principle of selective techniques, developed particularly by Lederberg in his bacterial studies, greatly simplified the detection of rare events. Benzer's application of selective techniques permitted the normal (r+) phage, but not the rII phage to survive on strain K. Successful recombination between alleles would be recognized by the presence of r+ plaques on the E. coli K covering the agar. The "cross" between two different phage alleles was accomplished by simultaneous multiple infections of the host cells; those host cells which contained both types of infecting phage could be considered simulated diploids.

The significance of this discovery extended beyond the debate of intragenic and intergenic recombination among pseudoalleles. Benzer recognized that genetic concepts could be translated into molecular terms. "This great sensitivity prompts the question of how closely the attainable resolution approaches the molecular limits of the genetic material. From the experiments of Hershey and Chase, it appears practically certain that the genetic information of phage is carried in its DNA. . . . We wish to translate linkage distances, as derived from genetic recombination experiments into molecular units."[38] The molecular weight of DNA in one phage particle corresponds to that of 200,000 nucleotide pairs on a Watson-Crick model. The total genetic map, in Benzer's 1955 calculation, was about 200 map units in length. If one assumes a uniformity of crossing over, and the likelihood that all the phage DNA was haploid and functional, the genetic and molecular correspondence could then be made. "Given two phage mutants whose mutations are localized in their chromosomes at sites only one nucleotide pair apart, a cross between these mutants should give rise to a progeny population in which one particle in 10^5 results from recombination *between* the mutations (provided of course, that recombination is possible between adjacent nucleotide pairs)."[38]

Benzer's mapping procedure in this first paper was based on the frequency of recombinations found in different crosses. Outside marker genes to locate the alleles with respect to the left or right directions of the map were not used. The precise distances and relations of one mutant to another were lacking, but in general "the greater the linkage distance between the mutations, the larger the number of plaques that appear. . . ."[38] The work was incomplete, but its implications were clear. "By extension of these experiments to still more closely linked mutations, one may hope to characterize, in molecular terms, the sizes of the ultimate units of genetic recombination, mutation and 'function.' Our preliminary results suggest that the chromosomal elements separable by recombination are not larger than the order of a dozen nucleotide pairs (as calculated from the smallest non-zero recombination value) and that mutations involve variable lengths which may extend over hundreds of nucleotide pairs."[38]

A more conservative view was expressed by Demerec that same year. "I believe that it is much too early to speculate about the chemical structure of genes, or about the nature of the genic changes that are responsible for the appearance of various alleles. I only wish to point out that the chemical structure of deoxyribonucleic acid postulated by Watson and Crick would account for the events at a locus which are detectable by genetic methods."[106] In another paper that year Demerec and his colleagues presented results similar to those found by Benzer, but their interpretation was not as daring. "It is evident that the results of transduction can be detected only if the donor and the recipient bacteria differ in a genetic constitution, and therefore transduction experiments can be used to determine whether or not two similar mutants are genetically identical. For example, transduction between two cystineless strains would not be detectable if the deficiencies they carried were due to modifications at identical sites on their chromosomes, but it would be detectable if these modifications were not identical. . . . When we began the first transduction tests to determine allelic relationships among 25 cystine-requiring mutants, we were greatly surprised by the results, which showed that transduction to wild type occurred in all but a very few combinations. It did not seem reasonable to suppose that almost twenty gene loci, each playing a major role in the synthesis of cystine, could exist in Salmonella."[107]

The implications of such widespread recombination among the cystine-requiring mutants were more likely to be found in a genic interpretation. "If this is so, then a gene locus extends over a section of a chromosome; changes occurring in different regions of this section give rise to different alleles; and the transduction observed between alleles is due to incorporation into the transduced chromosome of a normal, unchanged region brought by phage from the donor to the recipient bacterium."[107] As in Benzer's first description, a lot more was needed to be done. The actual location of the sites in the different cystine regions could not be determined and the calculated map distances were difficult to repeat. "Unfortunately, the efficiency of transduction is affected by several factors that we are not yet able to control, so that quantitative data obtained in different experiments are not always comparable and consequently the values for recombination between nonidentical alleles give only an approximate indication of the distances between these alleles and their order of arrangement within a locus."[107]

By 1956 Benzer had improved the resolution of his system by a more intensive use of recombinational techniques which provided more detailed and accurate distance relations than those which he had obtained in 1955. Moreover, he reported a more complex organization of the rII region; it contained two segments which were apparently different in function. He named these segments A and B. With a much larger number of mutants analyzed, the "fine structure of the gene" became more precise. "A signal achievement of genetics has been the demonstration that hereditary factors, the classical 'genes,' are ordered in a one dimensional array in chromosomes divisible by genetic recombination. It is fascinating to inquire whether this one dimensionality and divisibility persists down to the ultimate molecular structure of the genetic material."[39] Benzer felt that this was the case. "It seems feasible to arrive at clear characterizations of the ultimate units of physiological function, mutation, and gene recombination and to place an estimate upon the molecular size of each. . . . By combining the total linkage length and total DNA content of a T4 particle, a very rough conversion factor may be calculated by translating linkage distances into molecular ones. This ratio is of the order of 10^{-3} percent recombination per nucleotide pair. According to this value, the functional segments of the rII region are about 4000 nucleotide pairs long. . . . The chromosome would seem to be divisible by recombination at least to the extent of pieces ten nucleotide pairs in length. The site of a mutation, i.e., the length of chromosome in which alteration is apparent, varies from several hundred nucleotide pairs for the 'anomalous' mutants to no more than ten for others."[39] The "anomalous mutants" were presumed to be deletions or rearrangements. Benzer recognized the similarity of his findings with the predictions made by Pontecorvo for the high degree of resolution potentially detectable in fungi and other systems. "It may not be unreasonable to hope that the structure and function of the genetic material in phage, which is so accessible to detailed study, can serve as a model for higher forms."[39]

Demerec, also in 1956, extended his studies of transduction in *Salmonella* to the tryptophan region. To his surprise there were several functionally distinct regions that were closely linked to one another, forming a sequential series of genes. "The finding of correspondence between order of loci on the linkage map and sequence of biochemical blocks is the most unexpected result of this study, particularly since an unsuccessful

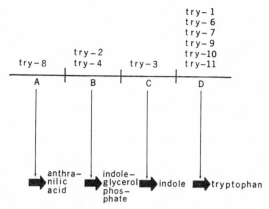

Figure 31. Demerec's Sequential Model of Gene Structure. In *Salmonella typhi-murium* there are four genes involved in the enzymatic synthesis of tryptophan. These genes A, B, C, and D have an internal fine structure. Each gene affects the step indicated by the heavy arrow. These linked genes were later found to be characteristic of bacterial operons. The tryptophan biochemical pathway is correlated to gene sequence in *Salmonella* but it is not correlated to gene sequence in fungi and other cellular organisms.

search for some such relationship had been made by Pontecorvo . . . working with fungi. There is very little doubt that such an organization, if it occurs at all in fungi and other higher organisms that have been studied genetically, is extremely rare; otherwise it would have been detected, especially in such well analyzed materials as Drosophila, maize, and Neurospora. In fact, there is some evidence that in Neurospora that certain loci which appear to be homologous to those studied by us are neither adjacent to one another nor linked in the sequence of their biochemical blocks."[108] This suggested a model similar to Pontecorvo's "millimicromolar" sequential system. Alternatively, "another way of explaining the observed difference between bacteria and other groups of organisms is on the assumption that in bacteria the nucleus is the site of metabolism and that all metabolic reactions are carried on by the genes themselves, whereas in higher organisms these reactions are performed by some organelles in the cytosome."[108] It is interesting that Lewis's model of linked sequential genes for pseudoalleles was not cited in these discussions of the linked tryptophan series. Probably Pontecorvo's use of fungi and a biochemical approach made his system more directly analogous to that found in *Salmonella*. The existence of two functionally different but adjacent members of the rII region was more closely analogous to Lewis's system than to Pontecorvo's because the phenotype of the rII region was morphological rather than biochemical in nature.

While the existence of fine structure in both bacteria and viruses was unambiguous, the attempt to detect such fine structure in the lozenge region of *Drosophila* by M. M. and K. C. Green was a failure. "The re-combination analysis of several *lz* mutants reported herein failed to un-cover any loci in addition to the three described previously (. . . 1949). This observation, taken at face value, is at variance with the observations of Benzer . . . in bacteriophage, Demerec, et al. . . . in Salmonella, and Pritchard . . . in Aspergillus. . . . The mutants are not, recombinationally speaking, continuously distributed, and failure to observe recombination supports the interpretation of allelism of the mutants. . . . Since the de-tailed analysis of the *lz* mutants shows that they are classifiable into but three loci, the model of the functional gene as being distinct from a re-combinational gene is less likely than its alternative."[159]

Also in 1956 Pontecorvo reasserted his hypothesis of functional over-lapping between genes. " 'Allelism' is a relation between elements of the genetic material which is approached asymptotically as we consider elements located nearer and nearer along the chromosome, or more and more similar in phenotypic effects. To approach this asymptote is the best that we can do at present because we have no basis for predicting from first princi-ples where the actual ultimate functional or structural unit of resolution will lie. . . . It seems quite possible that at least in some cases the transition between the sites of one system of integration, that is, one gene, and the next, may be gradual and with overlaps just as Muller and Goldschmidt suggest."[261]

In his third paper on fine structure of the rII region, in 1957, Benzer attempted a new conceptual definition of the gene, encompassing a tri-partite distinction based on the fine structure resolutions for mutation, function, and recombination. "The unit of recombination will be defined as the smallest element in the one-dimensional array that is interchangeable (but not divisible) by genetic recombination. One such element will be referred to as a 'recon.' The unit of mutation, the 'muton,' will be defined as the smallest element that, when altered, can give rise to a mutant form of organism. A unit of function is more difficult to define. . . . A functional unit can be defined genetically, independently of biochemical information, by means of the elegant cis-trans comparison devised by Lewis. It turns out that a group of non-complementary mutants falls within a limited segment of the genetic map. Such a map segment, corresponding to a function which is unitary as defined by the cis-trans test . . . will be referred to as a 'cistron.' "[40]

The estimates of the molecular weights of these three units depended on the same basic information as the two previous attempts to give nucleo-tide dimensions to the conceptual aspects of the gene. On the basis of a view later shown to be in error, Benzer assumed that only 40 percent of the total phage DNA was genetically active. He also boosted his map length from 200 units to 800 units by summing all his smaller lengths within fine structure regions and applying this to all loci in the phage. This effectively increased the number of nucleotides for his smallest units by almost one

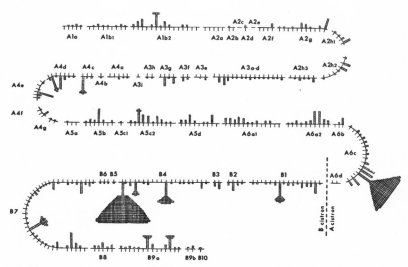

Figure 32. Benzer's Fine Structure Analysis of the rII Region. This virus T4 was used to generate mutations in a specific gene. The localization of these mutations involved numerous selective techniques. The rapid lysis region, rII, consists of two genes or cistrons. Each of these genes has an internal fine structure which Benzer correlated with nucleotide composition. The cistron usually consists of 100 or more sites or nucleotide pairs. Each site may be considered a nucleotide pair and the smallest distance between two sites is consistent with their molecular representation as two adjacent nucleotide pairs. This work on fine structure permitted the molecular analysis of mutation and the correlation of the gene with the Watson-Crick model of DNA. (After Benzer, 1961, Proc. Nat. Acad. Sci. 47:403–416.)

order of magnitude over those he had made the year before. There were, however, new features in this work which made it more significant than the previous two papers. In this new analysis, Benzer used his "anomalous" mutants or deletions to construct an overlapping deletion map, similar to the step-allele map of scute. In this case, however, the deletions were restricted to portions of the rII region and they did not consist of gross extragenic abnormalities such as those involved in the scute analysis. The localization of mutants to overlapping or non-overlapping regions of this deletion map enabled Benzer to check, independently, the localizations he made on the basis of frequency of recombinations. Also, the detection of a minimal frequency of recombination, between 10^{-2} and 10^{-3} percent established a limit to fine structure, or as Benzer expressed it, "it defines an 'absolute zero' for recombination probabilities."[40] In terms of map distance, the recon was 0.02 map units, the muton was 0.05 map units, and the cistron was several map units in length. In the A cistron Benzer had localized 60 sites for 241 of the 923 rII mutants he had available. Thus at least 240 sites were likely to be found when all the mutants were localized. "It seems safe to conclude that in the A cistron alone there are over a hundred 'sensitive' points, i.e., locations at which a mutational event leads to an observable phenotypic effect."[40]

In 1959 an attempt was made by Benzer to dispense with recombination frequencies altogether in attempting to prove the linearity of the sites in the rII region. "As pointed out by Muller in regard to similar difficulties encountered in mapping on the chromosomal level, strict additivity of 'distances' should not be taken as the criterion for the linear character of an array. A crucial examination of the questions should be made from the point of view of *topology* since it is a matter of how the parts of a structure are *connected* to each other, rather than of the distances between them. Experiments to explore the topology should ask qualitative questions (e.g. do two parts of the structure touch each other or not?) rather than quantitative ones (how far apart are they?)."[41]

There were two important consequences of the topological approach. First, it provided an accurate check on the recombinational mapping method; second, it eliminated Pontecorvo's asymptotic model of gene function. "How does the assignment of mutations into cistrons by the *functional* test correlate with their locations on the map derived from the independent recombination test? It turns out that the map may be split into two portions. . . . All mutants assigned to group A by the functional test have mutations to the *left* of the divide on the basis of recombination. All mutations causing functional defect B lie to the *right* of the divide. Those few mutants which are defective in both functions have mutations extending to both sides. Thus each cistron corresponds to a sharply limited segment of the linear structure."[41]

Deletion mapping, however, does not provide fine structure in the detail that recombination mapping does, since the deletions are likely to involve larger segments of the map or corresponding sequences of nucleotides. "The particular mutants used in the present work represent a special class of the spontaneously-arising rII mutants, having been chosen for their non-reverting and non-leaky character. Mutants containing large structural alterations would naturally be found in this group, and such large mutations are necessary for overlaps of the sort needed to test the topology of the structure."[41] A second feature of the mapping process, recognized for several years in the fine structure analysis, was discussed in critical detail in 1961. "Inquiry is made into the topography of the structure, i.e., local differences in the properties of its parts. Specifically, are all the subelements equally mutable? If so, mutations should occur at random throughout the structure and the topography would be trivial. On the other hand, sites or regions of unusually high or low mutability would be interesting topographic features."[43] While such "interesting" features might be more exciting for research, they would hamper the critical analysis of the maximum number of sites in a cistron. This number can be calculated from the Poisson distributions of 1, 2, or more occurrences of allelism at a site. The assumed randomness of the allelism in the Poisson estimate requires the "trivial" organization of the topography. Despite the existence of "hot spots" in the map, Benzer estimated that the spontaneous map was about 70 percent saturated; this meant that there were approximately 400 sites in the rII region. This estimate was not likely to be raised by a full order of magnitude. "The physical structure corresponding to the rII region represents less than 1 percent of the total DNA of the phage particle, or less than 2000 nucleotide pairs. If this is so, the number of possible sites would be of the order of at least one-fifth of the number of nucleotide pairs."[43] Thus the cistron took on a smaller size than it had been estimated to have in 1956.

Several enigmatic features of the topography have not been clarified. In either technique, the cistron has been restricted to a class of mutants that can be called amorphs or non-leaky mutants. There are however abundant rII mutants which have partial function, but these are kept out of the analysis to permit a more rigid, unambiguous test for linearity. To what extent do these hypomorphic or leaky sites correspond to the stable sites? To what extent is the Poisson method valid for estimating the map size? Are the instances of 2, 3, or 4 allelic recurrences random or are they examples of "lukewarm" spots? Another problem is the size of the muton. This is, at best, an indirect inference. For recombination measures the distance between two elements; it does not measure the width of an element itself. Thus, if a, b, and c are three sites linked in that order, the distances ab plus bc may equal the distance ac but that reveals nothing about the size of b since it is a recombinational relation that would be found whether the elements were single nucleotides or blocks of material of considerable size within which recombination does not occur. Thus the muton when translated into molecular terms is reduced to the smallest interchangeable element of the map, a feature characteristic of the recon itself. Another problem encountered in the molecular interpretation of fine structure is the apparent "ultrafine structure" which Tessman obtained among the alleles of hot spots. He estimated that the recon would descend "below absolute zero" on Benzer's scale and that there may be no lower limit for recombination. All of these problems point out that there is still work to be done to achieve a precision of translation between genetic and molecular languages.

CHAPTER 23

Complementation: Maps, Patterns, and Units

White eyes was the first multiple allelic series obtained in *Drosophila*. As the number of alleles in this series increased between 1913 and 1920, all were found to show the same phenomenon in allelic compounds; the compound individual was phenotypically intermediate between the eye colors manifested by either allele alone. Another series, studied by Muller, did not obey this rule. "This is the case of the truncate series, in which I have found that different mutant genes at the same locus may cause either a shortening of the wing, an eruption on the thorax, a lethal effect, or any combination of two or three of these characters. In such a case we may be dealing either with changes of different types occurring in the same material or with changes (possibly quantitative changes, similar in type) occurring in different component parts of one gene."[233] Muller's analysis of this system of alleles in 1921, and subsequent years, demonstrated that the compound of certain pairs of alleles expressed a normal phenotype. This series, later renamed the dumpy series by Bridges, remained an enigmatic case of complementation between members of a multiple allelic system. If the oblique wing effect is symbolized by (o), the thoracic vortices by (v), and the lethality by (1), then the mutant compound o/1v would be wild-type but the compound o1/ov would manifest (o). In general, three alleles are required to establish complementation. If a, b, and c represent these alleles and if the pair a/c is phenotypically wild-type but the pairs a/b and b/c are mutant, then complementation is functionally defined. Although dumpy was the first case to show this for qualitatively different effects of a mutiple allelic series, it is not unique. The mutants

210

achaete and scute form a wild-type compound as do the alleles facet-notched and facet bristles, both being recessive members of the Notch series. However, in 1932 Muller reported another type of complementation. "We have to report exceptions of the opposite type . . . namely, those in which the compound is more like normal in respect to effects in which both allelomorphs are similarly abnormal. One such case is that of lozenge-eye in combination with a particular spectacled-eye allelomorph of it."[233]

Bentley Glass, who had discovered the complementation between facet-notched and facet, attempted to explain this and the dumpy case. "If we accept certain indications, not by any means to be regarded as certain proof, that truncate itself is a deficiency, it is possible that dumpy, the single other mutation including both the character affecting the wing and that producing vortices on the thorax, is not a single gene, but two neighboring ones, produced by a double mutation. But in view of the regular behaviour of dumpy, and the failure to detect crossing-over between the alleged parts in many years of use by many workers, it seems on the whole hardly probable."[138] It is interesting that Sturtevant and Morgan, in their analysis of eosin, assumed that white was a double mutant and attempted to extract eosin from females heterozygous for red and white. In Glass's case, the assumption that o and v were adjacent genes implied that ol and lv were deletions of either of these genes and olv was a deletion of both. Thus ov (dumpy) would be a double mutant if the assumptions on a "presence and absence" model were correct. This model would favor testing dumpy and wild-type for recombination rather than testing the compounds of dumpy alleles (save for the compound o/v). Had the test with o/v been tried and been found successful it is not likely that pseudoallelism would have been the immediate interpretation or that intragenic recombination would have been suspected. It would have been assumed, instead, that a presence and absence model had been demonstrated for the truncate system!

No further interpretations of pseudoallelism were attempted until Lewis studied the complementing alleles of the bithorax system; he suggested that the complementation occurred because the blocked intermediate from one chromosome could diffuse to the chromosome bearing the other allele. Disruption of this diffusion by rearrangements in either chromosome presumably forced apart these somatically synapsed chromosomes and prevented the diffusion of their products. This "transvection" effect constituted a type of position effect in Lewis's system which could accommodate allelic complementation. Pontecorvo did not interpret complementation in this fashion. He did, however, assume some sort of differences in the diffusion of gene products. These differences were based on the linear distance separating the mutant sites localized on the genetic map. The closer the localizations of two allelic sites, the less the complementation between them. This "asymptotic" model was abandoned the following year with the independent discoveries by two groups of investigators of complementation in cells of Neurospora.

In Aspergillus, Pontecorvo had used cells whose nuclei were diploid. But in Neurospora the cells contained two nuclei, each haploid, and the complementation between these alleles thus had to involve an event in the cytoplasm of the cells. Fincham and Pateman in England used two alleles of the glutamic dehydrogenase region. These two alleles, am^{32} and am^{47} were extremely closely linked. "It is clear that am^{32} and am^{47} heterocaryons are able to produce glutamic dehydrogenase, though at a level considerably less than the wild type. . . . These results might readily be explained if 'glutamic dehydrogenase' of Neurospora were really two enzymes, acting in sequence. This seems unlikely . . . and . . . virtually ruled out by experiments in which we have failed to detect activity either in mixtures of am^{32} and am^{47} extracts or in extracts of mixtures of am^{32} and am^{47} mycelium. . . . It may be supposed that the synthesis of the enzyme includes two steps mediated by two distinct nuclear products. . . . Alternatively, one may imagine that the wild type am locus acts as a unit in producing a single component of the enzyme forming system and that this component can be partially reconstituted in the cytoplasm of the heterocaryon from two different defective nuclear products."[128]

In the United States, N. H. Giles and his students, Partridge and Nelson, carried out an extensive study of the adenylosuccinase alleles. "A series of over twenty adenine-requiring mutants of independent origin has been obtained in one wild-type strain as a result of mutation at a single locus which is quite favorable for genetical studies, having suitable closely linked markers on either side. These mutants are all characterized by being deficient for a single enzyme involved in the terminal step of adenine biosynthesis—adenylosuccinase—for which a convenient assay procedure is available. . . . It has been possible to test for heterocaryon formation and growth in the absence of adenine with the various strains. The unexpected result has been that certain combinations of mutants are able to form adenine-independent heterocaryons . . . in which appreciable adenylosuccinase activity can be demonstrated, even though both components lack detectable adenylosuccinase activity."[137] Giles, like Fincham, ruled out the possibility of two separate reactions; he also was able to exclude reversion because the haploid nuclei could be extracted in the subsequent generation and each nucleus retained the mutant characteristic. "A reasonable possibility would seem to be that, in cases where complementation occurs, the two mutants involved have arisen as a result of mutational changes at different sites. . . . Where complementation fails, the two mutants may have a certain portion of the locus damaged in common."[137] While Giles did not take a strong stand on any mechanism for complementation, he recognized a feature of the system that was important for the gene concept. "The present results appear to make more difficult the general application of the cis-trans position effect test to delimit a locus as a functional unit . . .

at least if functional unity is defined as control over the synthesis of a single enzyme."[137] The cistron, based on the rII analysis, was applicable to the adenylosuccinase region only if complementing mutants were excluded, even if such mutants, as homocaryons, were non-leaky and stable!

The extension of complementation to the analysis of the gene was first attempted in the step-allele theory developed by Serebrovsky, Dubinin, and Agol. In their approach the intensity of complementation could be used for mapping the various mutants. Almost thirty years later the same idea independently occurred to D. G. Catcheside and A. Overton. In 1958 they proposed a method for mapping the various complementing mutants without employing recombination. "The two-dimensional pattern of reactions can be represented by a one-dimensional diagram, in which each mutant exerts a range of influence. . . . These ranges are drawn in such a way that overlapping ranges occur where the heterocaryons do not grow on minimal medium, and non-overlapping ranges where complementation occurs in the heterocaryons. A completely consistent representation of the data is obtained in this way. The problem now arises as to whether this linear diagram has anything more than a formal meaning. That the diagram is linear is highly suggestive that the data on which it is based disclose some kind of linear differentiation within the gene, or its product."[85]

In the arginine region (arg-1) examined by Catcheside and Overton, four sub-units of this map were inferred. If these sub-units are designated 1, 2, 3, and 4, the mutants of pattern A affected all four sub-units; B affected 1; C affected 1, 2, and 3; D affected 3 and 4; E affected 4; and F affected 2. "The diagram could be interpreted as indicating that the arg-1 locus is complex and made up of four physiologically distinct blocks, say four cistrons in Benzer's sense. . . . On this interpretation, the members of group A would have to be regarded as deletions encompassing all four cistrons. . . ." This was rejected, however, because one of the mutants of the A pattern was capable of reverting to wild-type. "The interpretation . . . encounters difficulties which render it untenable." Catcheside was not able to obtain recombinations in his system to make a comparison between the two maps. "It must be remembered that the interaction occurs in a heterocaryon and therefore that the site of the interaction is probably in the cytoplasm. . . . Hence, the interaction must be between gene products, either at the level of the formation of the units from which potential enzymes are built or at the level of the final shaping of the enzymes."[85] That same year M. Case, working in Giles's laboratory, also found that a cistron interpretation of each sub-unit of the complementation map was untenable. In their analysis, Case and Giles used both recombination and reverse mutation to test the possibility of deletion for some of the non-complementing mutants. "Neither the recombination data nor the reversion data support the view that such non-complementing types carry defects in both cistrons, as for example, would be expected if they had resulted from deletions including adjacent regions from both cistrons. The explanation for this large class of completely non-complementing mutants is not yet clear."[61]

Giles also reported that his student D. Woodward had found that a linear map could be derived from the complementation patterns of the adenylosuccinase region (ad-4). "Based on the complementation pattern, it is possible to arrange these ad-4 alleles in a linear sequence. This can be done because of apparent overlap between non-functional regions characteristic of the individual alleles. The relationships of the 15 different complementing types obtained indicate a total of seven distinct functional units (cistrons) in this 'complementation map' of the ad-4 locus."[136]

Woodward, however, had a more significant observation to report on his findings in 1959. "By mixing mycelial homogenates from certain pairs of complementing adenylosuccinaseless Neurospora mutants grown separately, adenylosuccinase activity as high as 20 percent of that obtained from in vivo complementation has been recovered. . . . Additional data have also been obtained from farther studies involving in vivo restoration of enzyme activity. These data support earlier findings of a correlation between complementation map distance and the levels of adenylosuccinase activity found in interallelic heterocaryons."[320] In Woodward's analysis the complementation map was extended to eight complementation units, or "cistrons" as he designated them. "The results . . . indicate a steady increase of enzyme activity with increasing map distance up to a separation of about 4 cistrons, at which point no further significant increase in activity is observed. . . . If the sites of damage in the two different defective nuclei are sufficiently widely separated such that random exchange can occur, then such a heterocaryon should yield 25 percent of wild-type enzyme activity. This theoretical maximum of 25 percent was in fact the maximum adenylosuccinase yield observed. . . . Until more is understood about in vitro complementation, definitive statements concerning the mechanism involved cannot be made. Nevertheless, the present system appears to provide an excellent opportunity for studying this problem."[320]

Neither Fincham nor Giles attempted to relate their models of complementation to those of Lewis or Pontecorvo. Both Lewis and Pontecorvo proposed their theories of complementation independently in 1955. Either Fincham and Giles did not see the relation of their own work to the phenomenon Lewis and Pontecorvo described, or they considered that their work, which was based on the activity of a demonstrable enzyme in a heterocaryotic system, was completely novel and unrelated in principle to the phenomena described by Lewis and Pontecorvo. Similarly, none of the contributors to complementation mapping cited the work on stepallelism or its implications. It is easy to understand how the papers of Sturtevant and Schultz and that of Child could be overlooked in working out the nature of complementation maps, but it is difficult to see why the procedures and principles of the mapping in scute would be overlooked. It is possible that geneticists of the post-World War II generation have never encountered these alternative theories, but it is more likely that there was a genuine change in the habits of geneticists in this period. The intense competitive research characteristic of the 1950's was accompanied by a search for new techniques and the use of new organisms. The immense volume of research with these new organisms has led to a drifting apart of the different schools of genetics with subsequent loss of communication and interest in each other's work.

In the midst of this period of interest in complementation, I was completing my Ph.D. studies on the dumpy locus. In choosing this system, I wish I could credit insight or historical curiosity. It was neither; I happened to find a crossover between *o* and *lv* from an *o/lv* female carrying an appropriate marker in one of the chromosomes. The cross was not designed for a pseudoallelic study so much as it was designed to illustrate complementation and allelism in a multiple allelic system. Furthermore the cross was not part of an experimental search, but part of a class exercise in Muller's advanced genetics laboratory. My only contribution to the primary event in this "accident" was to follow Bateson's admonition—I "treasured my exception," and it was passed on to Muller, who confirmed by recombinational tests that the exceptional fly was composed of *o* and *lv* in that order. That chance event, in 1955, of course led to a curiosity about the problems of pseudoallelism and complementation. When I discovered a crossover between *olv* and *ov* (again, unexpected; the cross was designed for another purpose which I have forgotten) the prior awareness of the problems cited by Glass and Muller on the one hand, and Pontecorvo and Lewis on the other, immediately forced my attention to the significance of this event. It eliminated the "presence and absence" type model of two adjacent units, *o* and *v*; it eliminated a model based on the simple combination of three units, *o*, *l*, and *v*; it made Lewis's model of separate genes for separate functions extremely unlikely in its simplest form; and I was stimulated to test the one remaining hypothesis: Pontecorvo's model of asymptotic complementation. The distribution of the mutants showed no relation of complementation to the distance apart of the mutants on the genetic map. This left the dumpy series without a prevailing model to account for its structure and function.[58] Complementation mapping had not been introduced at the time I prepared my thesis. When I attended the Cold Spring Harbor meetings where complementation mapping made its debut, I was surprised at the general agreement that it would be found to be colinear with genetic mapping. If this were so, the colinearity was not universal or the genetic systems of *Neurospora* and *Drosophila* were unrelated in the mechanisms of their complementation.

Case and Giles in 1960 extended their analysis of complementation to other loci. "It is now clear that the patterns of interallelic complementation at certain loci can be described in terms of linear (one dimensional) sequences of complementing mutants which have been designated *complementation maps*. This discovery of a non-genetic functional test which established a linear complementation map may reflect the linear organization of both a gene and its product. . . . At first glance there appears to be, in general, a rather striking agreement between the ordering of the mutants on the two maps. . . . Significant exceptions occur, indicating that strict colinearity does not hold."[62] The postulated exceptions to colinearity might involve mutants which caused folding or unfolding of the polypeptide produced. "The marked colinearity between the left ends of the two maps is quite striking and could reflect an absence of pronounced folding in one of the terminal regions of this particular enzyme protein."[62] They also modified their previous model on the basis of this detailed analysis of colinearity. "Clearly, present results of interallelic complementation studies in Neurospora demonstrate that considerable complexities exist. Hence attempts to classify functional groupings of related . . . mutants in terms of 'cistrons' appear premature. In fact the overall evidence appears to indicate that the application of the cis-trans test in defining a 'cistron' requires certain qualifications and restrictions."[62]

Later that same year I made a survey of all published work on complementation with the hope of understanding what a complementation unit represented. The non-linearity of the dumpy series suggested that other geometrical models were possible or that the complementation map itself was virtually meaningless. If linear maps did not correspond to anything, what constituted their linearity? This led me to generate a mathematical treatment of the linear maps published. In most cases the number

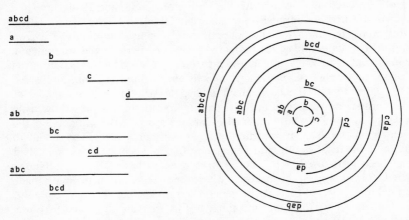

Figure 33. Linear and Circular Complementation Maps. The number of complementation patterns (line segments) in a linear map is smaller than that obtained in a circular map. For the four complementation units—a, b, c, d—patterns ac, bd, ad, abd, and cda are missing. Patterns ac and bd are missing in the circular pattern. A linear map has $n \left(\dfrac{n+1}{2} \right)$ patterns; a circular map has $n^2 - n + 1$ patterns; the maximum patterns in a three-dimensional model would be $n^2 - 1$. (After Carlson, 1961, Nature 191:788–790.)

of patterns fell short of the number required for full agreement with linearity, but it left them consistent with linearity. In the dumpy case the patterns exceeded the mathematical limits of patterns for three complementation units. I generated a complementation map for dumpy in the form of a circle and found that it fitted a circular model. Furthermore all the linear maps were also consistent with the circular model since they fell within the mathematical limits of circular maps. So too for all higher orders of geometrical models, based on polygons and three-dimensional constructions. "Although it is not possible either to confirm or to invalidate the linearity of complementation mapping, the phenomenon of complementation is a real one and its value for understanding gene function, and indirectly, gene structure, is not denied. Nor does the limitation imposed by the concept of the complementation unit rule out the cytoplasmic reassociation of component polypeptides of an enzyme or other protein. Such demonstrated mechanisms of cytoplasmic complementation may be independent of the structural assumptions inherent in the complementation mapping of the particular gene involved. It seems, however, that the alterations in the gene molecule and the functional disturbances produced in its final product are more complex than can be accounted for on the assumption of mere colinearity between genetic recombination and genetic complementation."[60]

In 1962 S. R. Gross examined the complementation patterns of 158 mutants affecting a locus involved in the synthesis of leucine (leu-2). "The topographical projection of the complementation patterns is . . . that it is linear, continuous, and overlapping. The simplest geometrical representation is circular but no specific geometry other than linear continuity is implied."[160] Another case of a circular complementation map was reported by A. M. Kapuler and H. Bernstein in 1963. They examined Ishakawa's data of 130 complementing mutants of the adenylosuccinate synthetase region (ad-8) in Neurospora. "On the basis of a large number of complementation tests among these 130 mutants, Ishakawa constructed a linear complementation map containing three basic units. . . . On re-examining these data, the present authors found that the interactions of all 130 complementing mutants defective at the ad-8 locus could be represented without exception by a nine unit circular complementation map. . . . The sites of mutation for seventy-three of the complementing . . . mutants were ordered in a linear genetic map. . . . Among these seventy-three mutants there were fifteen individual patterns of complementation. . . . The genetic map was allowed to assume a configuration about the complementation map such that radial lines . . . (drawn from each site on the genetic map to the corresponding subgroup on the complementation map) . . . did not cross. By following this procedure it was found that the genetic map formed a simple spiral around the complementation map."[176] Kapuler and Bernstein also recognized a principle found by Child and by Sturtevant and Schultz in their analysis of the step-allele map. "It is common experience that as one scores with increased sensitivity for intragenic complementation, more positive responses occur, and the complementation map consequently changes. . . . It seems unreasonable to assume that a map embodying all positive responses obtained at high sensitivity is intrinsically more informative than a map based on all positive responses measured at a moderate sensitivity. It is thought likely that there is no 'ultimate' complementation map for a locus, but rather a unique and characteristic map at any given level of resolution."[176]

Concurrent with the developments in fine structure and complementation theory were advances in the nature of mutation and the genetic code. One of these contributions, arising chiefly out of theories proposed by Crick and Brenner, favored the ideas that there were two classes of point mutations.[48] One type, "missense," usually involved single amino acid substitutions in which the loss of enzyme activity was not accompanied by the loss of immunological activity (the "cross reacting material" of such mutations leading to their designation as crm-mutants). The other type of mutation, "nonsense" mutations, presumably prevented the "reading" of the genetic code, and consequently no immunological activity could be detected among this class of revertible mutations. Using the physiological technique of isolating mutants which could not grow at higher temperatures (37° to 42° C) but which could grow at 25° C, R. S. Edgar, G. H. Denhart, and R. H. Epstein obtained a class of temperature sensitive (ts) mutants. They also were able to obtain mutations which grew on a strain of E. coli K but not on E. coli B (amber or am mutants). When these mutations were mapped, a surprising relation was found. "Intragenic complementation occurs between many ts mutations in the same gene, but not between am and ts mutations located in the same gene. These results suggest, on the basis of comparable studies with other organisms, that am mutations are of the non-complementing type and are 'nonsense' mutations, while ts mutations are of the 'complementing' type and thus are 'missense' mutations."[122] This necessity for mutations which do not destroy the major topological features of the gene product was in agreement with the many independently proposed models of complementation as a cytoplasmic interaction of the sub-units or components of functional enzymes.

An extension of this view was made by Crick and L. Orgel in 1964. They accepted the general view that functional proteins were composed of sub-units or "monomers" (the functional assemblage they designated as a "multimer"). Furthermore, their main point that "complementation is usually due to the correction of the misfolding of one monomer (produced by mutation) by some unaltered part of the other monomer,"[91] was similar to the viewpoint of Case and Giles. "To achieve complementation the misfolded region must lie so that it can only be corrected by part of the homologous region (correctly folded) in one of the monomers. . . . It is easily seen that this is most likely to occur in the regions adjacent to the axes of rotation of the multimer. This basic prediction can easily be seen to lead to the general type of relationship observed in complementation. Mutants affecting the same region, near a rotational axis, will often be near together on the genetic map. Thus in many cases such mutants will not complement one another. The length of a 'segment' in a complementation map will be a rough measure of the length of the region misfolded by the mutants in that segment. If misfolding spreads along the length of the polypeptide chain there will be a general tendency for the genetic map and the complementation map to be co-linear, but there will be many exceptions. If it spreads not along the chain but to adjacent folded segments, the complementation map may remain linear but it may not be co-linear with the genetic map. Complicated misfolding may easily produce non-linear, circular, or spiral complementation maps. . . . We believe . . .

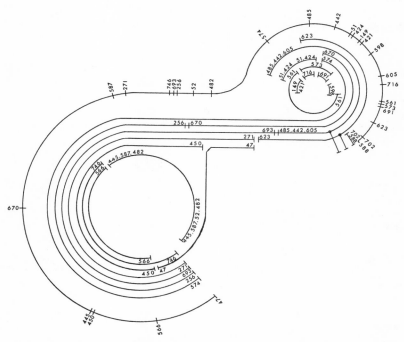

Figure 34. A Complex Complementation Map. This map of the ad$_8$ locus of *Schizomyces pombe* reveals how exceptional complementing alleles may distort the map from a simple linear or circular model. The structure of the map is no longer associated with the gene. Colinearity between sites on the gene (outer line) and the complementation map (inner lines) may reflect relative distance of mutational sites and amino acid replacements. The location of the replacement as well as the type of amino acid substitution are *both* significant in the three-dimensional structure of the protein sub-unit. Until adequate analyses of proteins from complementing systems have been made, the significance of complementation maps will remain unsettled. (After Leupold, V., and H. Gutz, 1965, Proc. XI Intnl. Congr. Genet., ed. S. J. Geerts, vol. 2:31–35.)

that the details of the folding of the polypeptide chain are the key to the relationship between the complementation map and the genetic map, but we do not believe that there is any *simple* general way of deducing these details from the supposed correspondence between the two maps."[91] This theory, however, could not be put to an immediate test. "Unfortunately, the structure of proteins is so complex that we have been unable to devise any crucial experimental test of our theory which does not depend on an a priori knowledge of the configuration of the protein. Such information is obtained only from lengthy X-ray studies. We believe that it is essential for any detailed test of our general theory or, indeed, of any other theory of complementation."[91]

This general tone of pessimism about the utility of complementation maps for revealing details of the structure of the polypeptide was also expressed by Catcheside in 1964. In some of his maps the linearity could be preserved by a hairpin folding, in others by spiral or irregular polygonal organization. "However, it cannot easily be supposed that the protein is merely a simple spiral or a hairpin. The clues provided by correlation of the genetic and complementation maps cannot be regarded as providing more than hints of the projection of the polypeptide chain on the surface of the monomer which, making contact with the homologous surfaces of other monomers, provides chances for correction of misfolding. Some of the complexities are no doubt due to the fact that the contact surface of one monomer may appose the contact surfaces of two or more other monomers in the same polymer."[84]

It is ironic that the step-allele and complementation theories have such remarkable parallel fates. Both initially hoped to order mutations by a functional rather than a recombinational test; both initially assumed a simple linear model of the constructed map; both assumed an initial colinearity between the gene and the map; both encountered exceptions to the map; both were found to be subject to reordering and repatterning when physiological changes were introduced; and both were reappraised in a pessimistic tone after their initial promise had become too tortured in complexity to provide a simple model. The future will provide many analyses of enzymes by the x-ray diffraction techniques developed by Kendrew, Perutz, and their colleagues. Whether these will provide definitive tests of complementation models is difficult to predict since the pattern of folding in crystalline form subjected to x-ray diffraction may not correspond to the type of folding existing among the monomers in the physiological states of complementation *in vivo* or *in vitro*.

CHAPTER 24

The Operon

When Bateson was a graduate student, it was a matter of faith that the problem of heredity would be solved through embryology. The German school of cytology developed this viewpoint theoretically with the introduction of particulate determiners in the chromosomes. Wilson popularized this view at the turn of the century and the chromosome theory that he and his students proposed soon found an experimental outlet in Morgan's laboratory. It is ironic that Morgan, who began and ended his career as an embryologist, shifted the direction of interest in heredity from the problem of development to the problems of the transmission of hereditary factors and the mechanics of Mendelism. The appeal of the theory of the gene, which Morgan widely popularized, was so strong that geneticists found little time to devote to developmental problems. Certain classes of physiological or developmental problems, however, were explored at the genic level, especially in the early 1940's. Thus lethal factors in *Drosophila* provided an opportunity for embryological studies; the control of biochemical pathways, as in the eye color system, was a second major field that found agreement with the gene theory. Nevertheless, the haunting question remained: how does variation occur in a clone of cells sharing identical genotypes?

Although this question was pushed aside in *Drosophila* research, either as a problem for the future or as one better studied by embryologists, it was the major impetus for studies in cytoplasmic inheritance, particularly in *Paramecium* and variegated plants. In addition to these two outlooks there was a phenomenon recognized by microbiologists but virtually unknown to geneticists. This was the physiological capacity of yeast cells to adapt to new chemical environments *en masse*, rather than through mutation and natural selection. If yeasts grown usually on sucrose were transferred to a lactose medium, they would grow poorly at first but after several hours they would be as efficient in metabolizing lactose as they had been on sucrose. The new metabolic functions were attributed to the production of enzymes which had not been present before. This phenomenon of "adaptive enzyme formation" was explored sporadically over a fifty-year period without satisfactory explanation. Its similarity to the ciliary antigen system, with only one antigen (or adaptive enzyme) dominating under any given condition, made it a candidate for the plasmagene theory. Later, when the plasmagene theory ran into difficulties, it remained an example of cytoplasmic inheritance.

In the late 1940's, studies of adaptive or induced enzyme synthesis in bacteria were carried out at the laboratory of J. Monod in Paris. The first breakthrough in the analysis of this problem was made in 1953 by Monod and G. Cohen-Bazire. They had found that two enzymes, β-galactosidase and tryptophan-desmolase responded in similar but opposite ways to the presence of specific metabolites. In the case of β-galactosidase, lactose and related compounds which have a β-galactoside would *induce* the formation of the enzyme to degrade it. In the case of tryptophan-desmolase, either tryptophan or its precursor, indole, would inhibit or *repress* the formation of the enzyme necessary for its synthesis. "These observations are narrowly comparable to those which we have reported for β-galactosidase, which is concerned with the kinetics of the phenomenon as well as the specificity relations between the enzyme and the inhibitors of its synthesis. The congruence of results bearing on such different enzymes encourages the hypothesis that these specific inhibitory effects express a general property of enzyme forming systems."[194] The most significant fact, however, was that "the biosynthesis of tryptophane-desmolase . . . is partly inhibited by tryptophane and by indole. . . . It seems that there is no inactivation of the enzyme at all, but only an effect on its synthesis."[194]

In the genetics of higher organisms, peculiar behavior was observed in maize at the nuclear level which seemed to defy the nearly universal properties of Mendelian genes. R. A. Brink in 1956 reported "a genetic change associated with the R locus in maize which is directed and potentially reversible."[55] Brink's observations were similar to those which led to the "contamination" hypothesis which Castle had proposed several decades earlier. The "integrity of the gene in heterozygotes, first inferred from Mendel's *Pisum* data and repeatedly confirmed since in a wide variety of organisms, is regarded as a universal principle in Mendelian heredity. . . . The present study is concerned with a seeming exception to the rule at the well-known R locus in maize conditioning aleurone and plant color."[55] Brink found that non-stippled ("self-colored") R^r/R^r plants when crossed as the pollen parent to colorless, r/r plants gave F_1 aleurones that were a dark mottled color, $R^r/r/r$. This simple dominance did not hold, however, when $R^r/r/r$ kernels were obtained from test crosses of r/r to heterozygotes bearing the stippled allele, R^r/R^{st}. Somehow the stippled allele converted or contaminated the R^r allele so that when crossed to r/r the R^r/r/r kernels were light mottled. Such pale R^r/r heterozygotes when crossed with each other would partially restore the normal color to the R^r/R^r homozygotes. Brink had no interpretation of this phenomenon, which was not a cytoplasmic phenomenon but somehow seemed to involve the control or regulation of the function of the R^r gene.

A more extensive study of regulatory or controlling elements in maize was made by B. McClintock. Her findings were unusual. "In maize, as in other organisms, a change at a particular locus in a chromosome is made evident by modification of a particular phenotype. . . . Because it is possible to predict which phenotypic character will be altered after modification at a particular known locus in a chromosome, it is inferred that some component is present there whose mode of action may be recognized. . . . These components appear to reside at fixed positions in the standard chromosome complement of maize, and they will be referred to as . . . the genes. Modification of action of a known gene component can result from the insertion of a controlling element at or near the locus of the gene; and, in general, the types of change in phenotypic expression induced by its presence there are those that could be anticipated from previously acquired knowledge of mutant expressions that have resulted from modifications at this locus. Controlling elements, on the other hand, need not occupy fixed positions in the chromosome complement, and detection of the presence of one such element depends upon characteristics it exhibits that are independent of its position."[189] These controlling elements caused changes in the mutation rate of genes in their vicinity as well as changes in the expression of such genes. "All the controlling elements so far identified . . . may have their area of activity confined within the nucleus itself, for they are known to serve as modifiers, suppressors, or inhibitors of gene action as well as mutators. They behave as if they were modulators of the genome. . . . Our present knowledge would suggest that gene elements and controlling elements represent two different classes of primary components of the chromosomes and that a close relationship exists between them."[189]

While these findings in maize directed attention to the nucleus as a possible source for the phenotypic regulation of gene activity in a clone, the distinction between the nuclear and the cytoplasmic outlooks was delineated by D. L. Nanney. Nanney, who had studied with Sonneborn, offered an interpretation of nucleo-cytoplasmic relations similar to the "dynamic" hypothesis frequently suggested by Goldschmidt. The chromosomal gene or "the 'master molecule' concept . . . presupposes a special type of material, distinct from the rest of the protoplasm, which directs the activities of the cell and functions as a reservoir of information. In its simplest form the concept places the 'master molecules' in the chromosomes and attributes the characteristics of an organism to their specific construction; all other cellular constituents are considered relatively inconsequential except as obedient servants of the masters. This is in essence the Theory of the Gene, interpreted as a totalitarian government. . . . (In) the 'Steady State' concept . . . we envision a dynamic self-perpetuating organization of a variety of molecular species which owes its specific properties not to the characteristics of any one kind of molecule, but to the functional interrelationships of the molecular species. Such a concept contains the notion of checks and balances in a system of biochemical reactions. In contrast to the totalitarian government by 'master molecules' the 'steady state' government is a more democratic organization, composed of interacting cellular fractions operating in self-perpetuating patterns."[252] This steady state model, based on the antigen system in *Paramecium*, accepted the role of specificity played by the nuclear genes, but attributed to the cytoplasm the regulatory mechanisms which fixed the alternative specificities according to the intra- or intercellular conditions. "It appears unlikely that the role of the genes in development is to be understood so long as the genes are considered as dictatorial elements in the cellular economy. It is not enough to know what a gene does *when* it manifests itself. One must know the mechanisms determining which of the many gene-controlled potentialities will be realized. In so far as these homeostatic mechanisms perpetuate a given specificity, they are 'hereditary' mechanisms, and their physical basis is a part of the physical basis of inheritance."[252]

Nanney's classification of the two outlooks made them seem to be more different than similar: the one, static and nuclear; the other, dynamic and cytoplasmic. This dichotomy made it difficult to predict the nucleo-cytoplasmic relation in each of these systems. In a comment on this paper, J. Lederberg revealed the nature of the impasse suggested by the two models. "I would be very much interested if anyone here could furnish the model for an experiment with which one could decide between the two models. I don't think that would be possible, because they are models of different kinds. The steady state is essentially a mathematical one, and the master molecule is essentially a political or particulate model, whatever you want to call it. They are neither exclusive nor non-exclusive—they are descriptions of different kinds of things that are going on. It is a lot like asking whether it is more correct to say that the George Washington Bridge is made of steel or whether the cables take the form of a catenary."[179] The impasse was actually surprisingly near to solution, and several apparently unrelated factors had already been discovered which, when pieced together, would provide the model that Lederberg did not think possible in 1957.

The first piece of evidence, enzyme induction and repression as a process of controlling the formation rather than the action of enzymes, was made in 1953 in Monod's laboratory. The second line of evidence appeared in 1955 with Demerec's discovery that the sequence of genetic factors controlling tryptophan synthesis in *Salmonella* was correlated with the biochemical sequence for tryptophan synthesis which S. Brenner had analyzed in Demerec's laboratory. The third line of evidence came from bacteriophage, although at the time of its discovery in 1956 its significance was not understood. E. Volkin and L. Astrachan at Oak Ridge, Tennessee, studied a small fraction of RNA that appeared to be participating in the metabolism of bacteriophage-infected cells. It had been found several years before that at the time of phage infection the DNA and RNA synthesis of the host is arrested. A few minutes after infection DNA synthesis begins again, but RNA synthesis remains at a low, barely perceptible, level. Using radioactive phosphorus to trace RNA metabolism in infected cells, Volkin and Astrachan confirmed the presence of a small portion of RNA which was synthesized after infection. They also isolated this RNA and found that its nucleotide base composition differed from that of RNA in uninfected bacterial cells. "Two possible processes, not mutually exclusive, may account for the observed heterogeneity in RNA-phosphorus uptake. One involves P^{32} incorporation into certain sections of the RNA molecule, not representative of the whole in terms of mononucleotide composition, while most of the molecule remains inert. The other process would involve synthesis of entire RNA molecules, but these must constitute a small percentage of the cell total RNA and must be a different mononucleotide composition from the composition of the whole. Such RNA molecules may be an entirely new species, possibly related to phage growth."[313]

Still another line of evidence that induction and repression were related to a genetic process was reported by A. Novick and M. Weiner in 1957. "The induced synthesis of β-galactosidase at low concentrations of inducer bears a close resemblance to the phenomenon of mutation (in the sense of a chromosomal change). In the case of mutation a cell is either mutant or wild type; in the case of enzyme induction a bacterium is either fully induced and makes β-galactosidase at maximum rate or is uninduced and makes no β-galactosidase. All the offspring of a mutant bacterium are mutants; all the progeny of an induced bacterium are induced, as long as maintenance inducer is present. However, unlike mutation, the induction system requires the continued presence of a low concentration of inducer to maintain the distinction between the induced and uninduced states."[254] Thus, enzyme induction was an all-or-none phenomenon with small quantities of an inducer-metabolite sufficient to "switch on" the production of a specific enzyme.

A "unified hypothesis" bringing biochemical approaches to a genetic problem of control and regulation was suggested by H. J. Vogel that same year. "Repression and induction are best pictured as control mechanisms rather than as inherent features of enzyme biogenesis. . . . The inducer would seem to act not by furnishing a prototype for the configuration of the enzyme molecule but, rather, by improving the performance of the template. Thus the inducer would assist the template catalyst and may be considered a 'promoter.' In contrast, a repressor may be compared to a catalyst 'poison.' It is proposed that inducers and repressors act by affecting the rate of dissociation of a template product from its template. Repression could then be the result of binding of newly formed enzyme protein to its site of synthesis through the agency of the repressor involved. Induction may reflect the neutralization of a binding effect which, in the absence of the inducer, would tend to hold the nascent protein near its template. . . . The hypothesis here proposed, which will be referred to as the 'regulator hypothesis,' implies that all enzymes, be they constitutive or adaptive, inducible or repressible, are synthesized in the same general manner but may differ from case to case in susceptibility to regulators."[312]

The final factor for the generation of an encompassing model of genetic regulation and control came from B. N. Ames and B. Barry in 1959. Hartman, who had participated with Demerec on the tryptophan and histidine regions of Salmonella, had turned over the biochemical analysis of this histidine region to Ames. The problem they posed was directly related to the structural association of different genes affecting a common pathway. "Does the histidine repression of enzyme synthesis affect each of the enzymes of the pathway to the same extent, i.e., is there a parallel increase in specific activity of all four enzymes on lowering the internal histidine pool and a constant ratio of the activity of one histidine enzyme to another independent of the concentration of the enzymes in the cell? . . . The data presented indicate that in acting as a repressor, histidine appears to affect the synthesis of each of the histidine biosynthetic enzymes to the same extent, within the experimental error of the assays. It is proposed that this phenomenon be called coordinate repression."[4]

The phenomenon of "coordinate repression" together with the earlier findings on adaptive enzyme formation, suggested a model: "one attractive possibility is that the repressor blocks enzyme synthesis at the gene level. The genes for the series of histidine enzymes have been shown by Hartman to be closely linked on the Salmonella chromosome and conceivably this length of the chromosome is 'functionally turned off' by a histidine–nucleic acid repressor (e.g., histidine-soluble RNA) with a specific affinity for part of the histidine section of the chromosome."[4]

This model, which was proposed in detail by F. Jacob and J. Monod, was analyzed genetically in the β-galactosidase system of E. coli. "We believe that these observations permit the conclusion that in the synthesis of many proteins, a dual genetic determination exists, with two functionally distinct genes participating: one of these (the structural gene) responsible for the structure of the molecule, the other (the regulatory gene) governing the expression of the first through the participation of a repressor. A particularly remarkable aspect of regulatory genes, additionally defining them, is their pleiotropic effect: in the cases so far known, the mutations which inactivate them simultaneously affect the synthesis of several proteins. This characteristic distinguishes regulatory genes from structural genes whose effects appear to be strictly limited to one protein, if not to one polypeptide sequence."[169] In their analysis, Jacob and Monod dis-

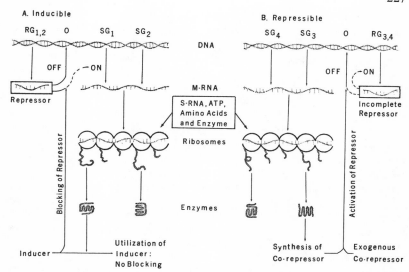

Figure 35. The Operon Model of Jacob and Monod. In an inducible enzyme system, A, the structural genes SG_1 and SG_2 do not transcribe their messages until a repressor (produced by a regulatory gene $RG_{1,2}$) is blocked. The repressor is blocked by an inducer, usually the compound on which the enzymes of SG_1 and SG_2 act. The transcribed message of SG_1 and SG_2 is messenger RNA (m-RNA). This message is decoded by ribosomes which move along the length of the m-RNA. Amino acids are brought in by energized soluble RNA (s-RNA). The decoded message is released as a peptide strand of increasing length until the end of the gene's message is reached. For repressible enzymes, B, the repressor produced by $RG_{3,4}$ is incomplete. Its completion is brought about by combination with a co-repressor, usually the compound synthesized by the enzymes of SG_3 and SG_4. Both inducible and repressible systems are feedback circuits in which excess of exogenous products causes changes in the genic control of enzyme synthesis.

covered a regulatory gene, *i* which apparently formed a repressor that could turn off the structural genes for β-galactosidase (z) and galactoside-permease (y). Mutations at *i* would convert the cell into a permanently de-repressed state (constitutive enzyme formation). "These observations pose the problem of the chemical nature of the repressor and its site of action. No answers are yet available for these questions."[169]

Nevertheless, Jacob and Monod predicted the properties of that site of action. "The expression of the group of genes would be the consequence of a unique structure, sensitive to the repressor and controlling their activity. A mutation affecting that hypothetical structure which may be called the 'operator' of the group of genes which it controls, would lead to a loss of sensitivity to the repressors. Furthermore, in a diploid, their action would only be exerted on structural genes aligned in a *cis* position with respect to the operator."[169]

The following year, 1960, Jacob and Monod with C. Sanchez and D.

Perrin, obtained proof for mutations affecting a region next to the structural genes which had precisely those predicted characteristics. This operator region was sensitive to the repressor of the regulatory gene. When mutated, it would sometimes lose its affinity for repressor, leading to a *dominant* constitutive mutation with coordinated effect on the structural genes it controlled. In other mutations of the operator region, the transcription or "reading" of the structural genes was prevented. The loss of function of the structural genes made it appear as if a deletion had removed the genes from the chromosome, but recombination tests proved their presence, and the genetic defect was attributed to a mutation in the operator region. "The hypothesis of the operator implies that between the classical gene, a unit of independent biochemical function, and the whole chromosome, there exists an intermediate genetic organization. The latter comprises coordinated units of expression (the operons) consisting of an operator and the group of structural genes coordinated with it."[172]

The theory of the operon predicted that the immediate product of reading from the genes would be an RNA molecule of base composition similar to that of the gene itself. Such molecules would have a temporary existence within the cell in contrast to the gene (DNA) itself. This type of molecule appeared to be identical to the "metabolic" RNA found by Volkin and Astrachan five years before. Three laboratories deliberately searched for this molecule and they reported its discovery in 1961. In England, Brenner, Jacob, and M. Meselson detected it in bacteriophage-infected cells. At Harvard, J. D. Watson and his colleagues characterized this molecule in both uninfected and infected bacterial cells. And, at Illinois, the laboratory of S. Spiegelman found the same product by using procedures nearly identical to those used at Harvard. Brenner, Jacob, and Meselson designated this molecule "messenger RNA." "A large amount of evidence suggests that genetic information for protein structure is encoded in deoxyribonucleic acid while the actual assembling of amino acids into proteins occurs in cytoplasmic ribonucleic acid particles called ribosomes. The fact that proteins are not synthesized directly on genes demands the existence of an intermediate information carrier."[47] This intermediate information carrier was not the ribosome itself, however; "rather . . . ribosomes are non-specialized structures which receive genetic information from the gene in the form of unstable intermediate or 'messenger.' . . . We may summarize our findings as follows: (1) after phage infection no new ribosomes can be detected. (2) A new RNA with a relatively rapid turnover is synthesized after phage infection. This RNA, which has a base composition corresponding to that of the phage DNA, is added to preexisting ribosomes, from which it can be detached. . . . (3) Most, and perhaps all, protein synthesis in the infected cell occurs in preexisting ribosomes. . . . It is a prediction of the hypothesis that the messenger RNA should be a simple copy of the gene, and its nucleotide composition should therefore correspond to that of the DNA. This appears to be the case in phage-infected cells. . . ."[47]

By 1961 the operon theory dissolved the impasse between the cytoplasmic and nuclear outlooks. The model to distinguish between the two outlooks turned out, instead, to "incorporate" both; the nucleus served as a storehouse of information and controlling switches which were responsive to cytoplasmic "signals." The phenomena of development and differentiation were neither due to plasmagenes nor to hierarchical self-perpetuating cytoplasmic "steady states." The operon became an extreme example of a mechanistic system of circuits, feedbacks, and blueprints; with it the cell assumed the analogy of the industrial plant rather than the analogy of the totalitarian state or the less disciplined democratic society.

In inducible systems the repressor would ordinarily seal off the operator and thus prevent the transcription of its structural genes. When inducer (such as a galactoside bearing molecule) was introduced in the β-galactosidase system, the repressor and the inducer would form a complex which would be incapable of turning off the operator, and thus the enzyme for the degradation of the inducer would be transcribed and decoded from its structural gene. In the case of a repressible system, the repressor was assumed to be incomplete. Thus the structural genes would be transcribed and their enzymes would synthesize a metabolite. This metabolite (e.g., tryptophan or histidine in the two systems analyzed) would form a complex that would complete the repressor. The combination of repressor and co-repressor in this complex would seal off the operator and thus prevent the further formation of enzymes for the synthesis of the metabolite (or co-repressor).

This theory was extended to embryology by Jacob and Monod as a working model for differentiation. "The occurrence of inductive and repressive effects in tissues of higher organisms has been observed in many instances, although it has not proved possible so far to analyze any of these systems in detail. . . . It has repeatedly been pointed out that enzymatic adaptation, as studied in microorganisms, offers a valuable model for the interpretation of biochemical coordination within tissues and between organs in higher organisms. . . . The fundamental problem of chemical physiology and of embryology is to understand why tissue cells do not all express, all the time, all the potentialities inherent in their genome. The survival of the organism requires that many, and, in some tissues, most of these potentialities be unexpressed, that is to say, repressed."[170]

At first, the chemical nature of the repressor was thought to be RNA. "No positive evidence has been obtained as yet concerning the chemical identity of repressors, but since they are presumably primary products of the regulators, the assumption that they are polyribonucleotides appears as the most reasonable guess. The nature of the interaction between repressor and effector (inducer or repressing metabolite) is unknown."[171] This was a logical inference based on a model of the operator consisting of a single strand of DNA with the repressor molecule having a complementary nucleotide composition.

By 1963, however, there were serious difficulties with this view of the repressor. Jacob and Monod, with J.-P. Changeux, revised their model and suggested that the repressor was a protein. This protein had the remarkable property of changing its activity when associated with induced or co-repressor effectors. "It would appear . . . that certain proteins, acting at critical metabolic steps, are electively endowed with specific functions of regulation and coordination; through the agency of these proteins, a given biochemical reaction is eventually controlled by a metabolite acting apparently as a physiological 'signal' rather than as a chemically necessary component of the reaction itself. . . . These proteins are assumed to possess two, or at least two, stereospecifically different, non-overlapping receptor sites. One of these, the active site, binds the substrate and is responsible for the biological activity of the protein. The other, or allosteric site, is complementary to the structure of another metabolite, the allosteric effector, which it binds specifically and reversibly."[171] The consequence of such allosteric proteins would be (1) an immediate cessation, for repressible systems, of activity at the cytoplasmic level in the presence of an allosteric effector; (2) an independence of the effector and the substrate on which an enzyme acts. This would permit, for example, hormones to act as effectors on many different enzymes affecting diverse metabolic products. The most important feature of the allosteric protein, however, would be its role as a repressor in genetic regulation. "It was still considered likely not long ago that the repressor might be a polyribonucleotide. . . .This assumption, which did not by itself account for the repressor-inducer interaction . . . has met with further serious difficulties, while several lines of indirect experimental evidence suggest that the active product of a regulator gene is a protein, present in exceedingly minute amounts in cells."[171] Since the original model of the repressor assumed stereospecificity between the repressor and its effector, and since the i gene presumably produced a single product, it was consistent that this product could be an allosteric protein. "It is evident that all these properties are immediately accounted for if the repressor is an allosteric protein possessing two sites, one of which binds the operator, the other the (positive or negative) effector."[171]

The application of the operon theory to a variety of inducible and repressible enzyme systems in bacteria has been highly successful. Its application to higher levels of organization has not been achieved because of the lack of detailed experimental procedures similar to those that have been used in bacteria. Nevertheless, the model has stimulated a new direction of nucleo-cytoplasmic relations and a renewed hope that the regulatory mechanisms of embryonic differentiation and morphogenetic movements will find a solution at the genetic level.

CHAPTER 25

The Coding Problem

For nearly sixty years geneticists have found the structure and function of the gene frequently beyond their experimental reach but nevertheless a subject of strong appeal. In 1906, Bateson expressed this longing for knowledge in eloquent terms. "But ever in our thoughts the question rings, what are these units that bring all this to pass which in their orderly distributions decide so many and perhaps all of the attributes and faculties of each creature before it is launched into separate existence? Colour, shape, habit, power of resistance to disease, and many another property that might be named, have one by one been analyzed and shown to be alike in the laws of their transmission, owing their excitation or extinction to . . . such units or factors. Upon them the success or failure of every living thing depends. How the pack is shuffled and dealt we begin to perceive: but what are they—the cards? Wild and inscrutable the question sounds, but genetic research may answer it yet."[15]

It was not possible in the first few decades of genetic research to do more than focus attention on the gross structure and location of the genes. The function of the genes was first expressed in a workable model by Beadle and Tatum with the one gene–one enzyme theory. In the 1940's there was sufficient interest in genetics to arouse the attention of physicists and biochemists. One of the first insights into this problem, based on intuition rather than experimentation, was made by the physicist E. Schrödinger in his book *What is Life?* According to Schrödinger, "the most essential part of a living cell—the chromosome fiber—may suitably be called an *aperiodic crystal*. In physics we have dealt hitherto only with *periodic crystals*. To a humble physicist's mind, these are very interesting and complicated objects; they constitute one of the most fascinating and complex material structures by which inanimate nature puzzles his wits.

Yet, compared with the aperiodic crystal, these are rather plain and dull."[273] This "aperiodic crystal" constituting the chromosome somehow controlled the entirety of an organism's specificity, structure, and uniqueness. "It is these chromosomes, or probably only an axial skeleton core of what we actually see under the microscope as the chromosome, that contain in some kind of code-script the entire pattern of the individual's future development and of its functioning in the mature state. Every complete set of chromosomes contains the full code. . . . In calling the structure of the chromosome fibers a code-script we mean that the all-penetrating mind, once conceived by Laplace, to which every causal connection lay immediately open, could tell from their structure whether the egg would develope, under suitable conditions, into a black cock or into a speckled hen, into a fly or a maize plant, a rhododendron, a beetle, a mouse, or a woman."[273]

In 1953 the model of DNA structure suggested by Watson and Crick explicitly included the concept of aperiodicity in the sequence of its nucleotide base pairs. "It . . . seems likely that the precise sequence of the bases is the code which carries the genetical information."[316] The bases adenine, guanine, cytosine, and thymine somehow controlled the specificity of proteins. Yet these chemicals were completely different from one another. An attempt to generate a genetic code for the translation of proteins from DNA was made in 1954. Surprisingly, it was suggested by an astronomer, G. Gamow, whose usual speculative forte was cosmogony! "The hereditary properties of any given organism could be characterized by a long number written in a four-digital system. On the other hand the enzymes (proteins), the composition of which must be completely determined by the deoxyribonucleic acid molecule, are long peptide chains formed by about twenty different kinds of amino acids, and can be considered as long 'words' based on a 20-letter alphabet. Thus the question arises about the way in which four-digital numbers can be translated into such 'words.' "[132]

Gamow's approach was direct. He assumed that the Watson-Crick duplex was the physical basis for the translation process. Within each set of four nucleotides forming a diamond-shaped hole in the surface of the duplex, a specific amino acid would drop into place. If the four nucleotides could form all possible patterns there would have been $4^4 = 256$ possible "diamond" patterns. But the DNA base arrangement was not random since the pairs A and T and G and C were non-randomly aligned by base-pairing. Gamow proposed, therefore, that only three of the four bases in a diamond were significant for specifying one amino acid. This would still have provided sixty-four possible combinations if the arrangements were random, but there was an overlap of the bases of any one diamond with those of the immediate diamond to its left or to its right. These restrictions limited the number of diamonds or "holes" in the DNA molecule to twenty different kinds, precisely the number to be found in the amino acid "alphabet." Using this supposition, Gamow argued that "free amino acids from the surrounding medium get caught into the

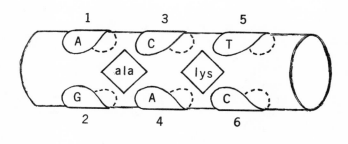

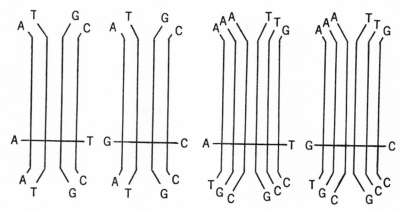

Figure 36. Gamow's Overlapping Triplet Code. The first genetic code assumed the protein was formed directly on the DNA molecule by fitting into the slots (diamond-shaped in the diagram) formed by base pairs. Only twenty possible neighboring bases can be formed. For the base pair 2 and 3 there are only four possible pairs. But 1 and 4 not being fixed by 2 and 3, have sixteen possible combinations. This code was proved incorrect by Brenner.

'holes' of deoxyribonucleic acid molecules, and thus unite into the corresponding peptide chains. . . . If this is true, there must exist a partial correlation between the neighboring amino acids in a protein molecule, since the neighboring holes have two common nucleotides.[132]"

Using all available amino acid sequences in proteins that had been degraded by numerous biochemists, S. Brenner came to a surprising conclusion. The overlapping code proposed by Gamow was impossible. If the sequence ABCD represented an overlapping sequence, then it could encode two amino acids. But the arrangement of the four bases could be made in $4^4 = 256$ ways. These 256 different dipeptide combinations were a little more than half of the theoretical number of dipeptides, which would be $20^2 = 400$ since there are twenty "letters" in the amino acid alphabet. Proving this, however, would have required an immense quantity of dipeptide sequences. Nevertheless, the impossibility of the overlapping code could be made from a tripeptide analysis! In the sequence ABCDE, five nucleotides would specify three amino acids. If the amino acid j is ABC, and k is BCD, then l would be CDE. Using the triplet for amino acid k, BCD, there would be only four possible left-most amino acid neighbors since only A would be "free" for a nucleotide not already determined by the overlap. Similarly, the amino acid k would share two of its nucleotides with amino acid l; hence l would also only have one free nucleotide (E) for four possible right-most amino acid neighbors. The number of actual left and right neighbors of amino acids exceeded this theoretical limit, forcing Brenner to reject either Gamow's overlapping code or the triplet sequence as the minimum coding unit for an amino acid. "The result has one important physical implication, the original formulation of overlapping codes was based on the similarity of the internucleotide distance in DNA to the spacing between amino acid residues in an extended polypeptide chain. It was supposed that each amino acid was spatially related in a one-to-one way with each nucleotide on a nucleic acid template. The present result shows that this cannot be so and that each amino acid is stereo-chemically related to at least two, if not three, nucleotides, depending on whether coding is partially overlapping or non-overlapping. . . . The non-overlapping of triplets implies that there must be some way of determining which triplets in a sequence are coding triplets and which are not. . . ."[47]

Although the Watson-Crick model was the main incentive for a study of the coding problem, there was strong support for a genetic code from the studies of protein chemistry. In 1949, L. Pauling, with the assistance of Itano, Singer, and Wells, took an interest in sickle-cell anemia. This disease had been recognized for some time as a genetic disorder. Its victims showed a sickle-shaped red blood cell (erythrocyte). Pauling suspected that the morphology of the cell was a secondary feature resulting from the semi-crystalline condensation of the hemoglobin in the erythrocytes of such individuals. Although a percent of American Negroes had the sickling trait in their blood, only one out of forty of such sickle-celled individuals suffered the disease in a severe form leading to their death in early childhood. Conventional chemical tests showed no differences between the two hemoglobins except for a reduced oxygen uptake by the sickle-cell hemoglobin. However, "a significant difference exists between the electrophoretic mobilities of hemoglobin derived from erythrocytes of normal individuals and those of sickle cell anemic individuals."[257] A careful study of the porphyrin molecules from the hemoglobins of normal and sickle-cell individuals showed no difference. "It is accordingly probable that normal and sickle cell anemia hemoglobin have different globins."[257] It was this difference in the protein portion of the hemoglobin molecule that caused the formation of partial alignments in the cells. These masses of aligned hemoglobin resulted in the distortion of the cell membrane giving the typical sickled shape. The various pathological defects could then be predicted from the primary defect in a hemoglobin molecule, according to Pauling's model. "If it is correct, it supplies a direct link between the existence of 'defective' hemoglobin and the pathological consequence of sickle cell disease,"[257] Such a condition Pauling called a "molecular disease."

Some seven years later, V. M. Ingram at the famed Cavendish laboratories in Cambridge devised a technique for characterizing the damage to the hemoglobin molecule in the sickle cell. "A new and rapid technique of characterizing the chemical properties of a protein in considerable detail has been devised; by its application a specific difference is found in the sequence of amino-acid residues of normal and sickle-cell hemoglobin. This difference appears to be confined to one small section of one of the poly-peptide chains."[168] Hemoglobin had earlier been found to consist of two chains, α and β. A functional hemoglobin molecule consisted of two α and two β chains as well as the porphyrin ring compound, heme, which was necessary for the association and dissociation of oxygen. The functional hemoglobin molecule contains about 600 amino acids. The specificity for a complete α or β chain, therefore was about 150. Normal hemoglobin was designated hemoglobin A and sickle-cell hemoglobin was designated hemoglobin S.

To approach this problem, Ingram used the enzyme trypsin to partially digest the hemoglobin molecules. "The action of trypsin on proteins is at present the most reliable way of splitting a peptide chain at specific peptide bonds. The enzyme attacks only those bonds which are derived from the carboxyl group of the amino acids lysine and arginine. There are about sixty of these in the hemoglobin A and S molecules, but since it is expected that each molecule is composed of two identical half-molecules, the number of peptides obtained by the action of trypsin should be about thirty, with an average chain length of ten amino acids. Small differences in the two proteins will result in small changes in one or more of these peptides. These should be detectable when the mixture is examined by a two-dimensional combination of paper electrophoresis and paper chromatography. It was decided to call the resulting chromatogram the 'fingerprint' of the protein."[168] The pattern of fingerprints obtained from hemoglobin A and from hemoglobin S were nearly identical. One peptide in hemoglobin A was missing on the chromatogram of hemoglobin S. In its stead was a fingerprint that did not exist in the normal hemoglobin. "One can now answer at least partly the question put earlier, and say that there is a difference in the amino acid in one small part of one of the polypeptide chains. This is particularly interesting in view of the genetic evidence that the formation of hemoglobin S is due to a mutation in a single gene. It remains to be seen exactly how large a portion of the chains is affected and how the sequences differ."[168]

The following year Ingram completed the peptide analysis of the two differing fingerprints. They differed by only one amino acid. A glutamic acid residue in the 'missing' peptide was replaced by a valine residue causing the peptide to appear in a different location in the hemoglobin S chromatogram. Thus out of some 300 amino acids, constituting the specificity for the α and β chains, only one amino acid was changed by the mutation. This change was serious enough to cause lethality when the defect was inherited from both parents.

A new approach to the coding problem emerged from several lines of evidence. The Watson-Crick theory specifically called attention to the inherent aperiodicity of DNA; the genetic fine structure of the bacteriophage chromosome was revealed by Benzer's analysis of the rII region; and numerous biochemical studies emphasized the important role of RNA in protein synthesis. In 1958 Crick presented a paper at the Society for Experimental Biology which ranks with Muller's 1921 address on the gene as one of the most stimulating and liberating conceptual papers in genetics. Although the major feature of Crick's article is the coding problem, its title "On Protein Synthesis" would erroneously lead astray most readers, who would infer that it is designed for biochemists. Crick's paper, like Muller's, does not present data; it suggests experiments and provides theories. "It is an essential feature of my argument that in biology proteins are uniquely important. They are not to be classified with polysaccharides, for example, which by comparison play a very minor role. Their nearest rivals are the nucleic acids. Watson said to me, a few years ago, 'The most significant thing about nucleic acids is that we don't know what they do.' By contrast, the most significant thing about proteins is that they can do almost anything."[89] Using the basic conclusion of Beadle and Tatum's analysis of *Neurospora*, Crick claimed that "the main function of proteins is to act as enzymes. . . . I shall also argue that the main function of the genetic material is to control (not necessarily directly) the synthesis of proteins. . . . Once the central and unique role of proteins is admitted there seems little point in genes doing anything else."[89]

The analysis of proteins by Sanger, duVigneaud, and others had shown that "a protein is to a large extent a linear molecule (in the topological sense) and there is little evidence that the backbone is ever branched. . . . A second important point—and I am surprised that it is not remarked more often—is that only about twenty different *kinds* of amino acids occur in proteins, and that these same twenty occur broadly speaking, in *all* proteins, of whatever origin—animal, plant or microorganism."[89] Furthermore, the specificity of proteins was long known (from immunological studies) to be unique for different species. Presumably this specificity resided in the genetic code that distinguished each species from one another. "Biologists should realize that before long we shall have a subject which might be called 'protein taxonomy'—the study of the amino acid sequences of the proteins of an organism and the comparison of them between species. It can be agreed that these sequences are the most delicate expression possible of the phenotype of an organism and that vast amounts of evolutionary information may be hidden away within them."[89]

Crick cited what might be called Sumner's paradox: if enzymes make everything in the cell, what makes enzymes? If the answer is "enzymes," then an infinite regression of enzymes would occur. "It is thus clear that the synthesis of proteins must be radically different from the synthesis of polysaccharides, lipids, co-enzymes, and other small molecules; that it must be relatively simple, and to a considerable extent uniform throughout Nature; that it must be highly specific, making few mistakes; and that in all probability it must be controlled at not too many removes by the genetic material in the organism."[89] The protein is a three-dimensional molecule, but it is essentially linear; thus the folding of the protein would have to be determined by the amino acids it contained rather than by the direct action of genes. Additionally, "the amino acids must be joined up in the right order. It is this problem, the problem of 'sequentialization,' which is the crux of the matter."[89] Crick suggested two hypotheses for a general coding of proteins by DNA. The first, or the sequence hypothesis, "assumes that the specificity of a piece of nucleic acid is expressed solely by the

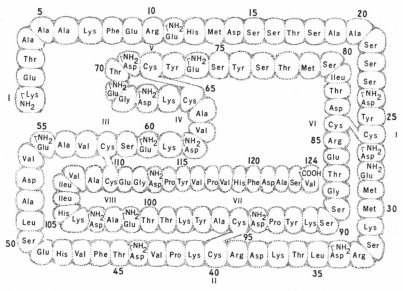

Figure 37. Linear Structure of the Amino Acid Sequence in Ribonuclease. The enzyme ribonuclease has been analyzed into its amino acid sequence. Its three-dimensional structure includes four disulfide linkages of cystine residues. The number 1 amino acid has a free amino group and the number 124 amino acid has a free carboxyl group. This molecule can be denatured to the linear form with loss of enzymatic activity and renatured to the three-dimensional form with resumption of enzymatic activity. (After Smyth, Stein, and Moore, 1963, J. Biol. Chem. 238:228.)

sequence of its bases, and that this sequence is a (simple) code for the amino acid sequence of a particular protein."[89] The second hypothesis, the central dogma, "states that once 'information' has passed into protein it cannot go out again. In more detail, the transfer of information from nucleic acid to nucleic acid, or from nucleic acid to protein may be possible, but transfer from protein to protein, or from protein to nucleic acid is impossible. Information means here the precise determination of sequence, either of bases in the nucleic acid or of amino acid residues in the protein."[89]

To read the genetic code from the DNA, Crick suggested that a complementary sequence of RNA is formed and enters the cytoplasm as a microsomal (or ribosomal) particle. In the cytoplasm, "adaptor" molecules carry the amino acid to the ribosomal RNA (also known as template RNA). "It is an essential feature of these ideas that there should be at least two types of RNA in the cytoplasm. The first, which we may call 'template RNA' is located inside the microsomal particles. It is probably synthesized in the nucleus . . . under the direction of DNA, and carries the information for sequentialization. . . . The other postulated type of RNA, which we may call 'metabolic RNA' is probably synthesized (from common intermediates) in the microsomal particles, where its sequence is determined by base pairing with the template RNA. Once outside the microsomal particles it becomes 'soluble RNA' and is constantly being broken down to form the common intermediates with the amino acids."[89]

Having worked out a scheme for the role of the genetic code in metabolism, Crick then turned to the nature of the code itself. The overlapping code had been proved to be wrong. Thus a non-overlapping code was necessary. But how would the "soluble RNA" recognize the proper triplets without some sort of punctuation? Since there were $4^3 = 64$ triplet combinations from the four bases, Crick assumed that 20 of these must be readable and the other 44 must be "nonsense." Using the general formula:

$$\text{A B} \begin{smallmatrix} A \\ B \end{smallmatrix} \quad \begin{smallmatrix} A \\ B \end{smallmatrix} C \begin{smallmatrix} A \\ B \\ C \end{smallmatrix} \quad \begin{smallmatrix} A \\ B \\ C \end{smallmatrix} D \begin{smallmatrix} A \\ B \\ C \\ D \end{smallmatrix}$$

the triplets making "sense" would be exactly twenty. Thus the sequence BCAADDCDBABBACC would be read as five consecutive amino acids since the triplets that overlap, such as CAA, AAD, DDC, would belong to the 44 nonsense triplets. Such a commaless, non-overlapping triplet code was also non-degenerate. Each amino acid was specified by only one unique triplet.

Later that year, however, a publication by A. N. Belozersky and A. S. Spirin of the U.S.S.R. weakened this optimistic approach to the coding problem. "The composition of deoxyribonucleic acid in different bacterial species belonging to various systematic groups was shown to vary widely (from the extreme AT-type to the highest GC-type); whereas that of ribonucleic acid from these species was shown to vary slightly, small differences being noted only between very diverse species. Thus the greater part of the ribonucleic acid of the cell appears to be independent of the deoxyribonucleic acid."[37] At best, only a small portion of the RNA seemed to have a base composition similar to that of DNA.

In 1959 Crick reported his feeling of frustration in a paper on "the present position of the coding problem." "The coding problem has so far passed through three phases. In the first, the vague phase, various suggestions were made, but none was sufficiently precise to admit disproof. The second phase, the optimistic phase, was initiated by Gamow in 1954, who was rash enough to suggest a fairly precise code. This stimulated a number of workers to show that his suggestions must be incorrect, and in doing so increased somewhat the precision of thinking in this field. The third phase, the confused phase, was initiated by the paper of Belozersky and Spirin in 1958. . . . The evidence presented there showed that our ideas were in some important respects too simple."[90] The possible alternatives that Crick listed were not exhaustive since the eventual solution of Belozersky's and Spirin's paradox was not among them. Crick suggested that "1) only part of the DNA codes protein. . . . 2) The DNA-to-RNA translation mechanism varies. . . . 3) The code is degenerate. . . . 4) The code is not universal. . . . 5) The nucleic acid code has less than four letters. . . . 6) The amino acid composition of the protein varies."[90] Crick did not like any of these alternatives and he felt that theory had gone as far as it was able to go and experimentation had to uncover new aspects before a fresh outlook could be established. However, "lest the reader be too discouraged, a few important experimental facts should be mentioned which any theory will have to explain: first, the evidence that the RNA of tobacco mosaic virus controls, at least in part, the amino acid sequence of the protein of the virus; second, the genetic effects of transforming principle, which appears to be pure DNA; and third, the genetic control of at least parts of the amino acid sequence of human hemoglobin."[90]

The newer approaches were already in a process of formulation. Using the rII region, E. Freese discovered different mutation processes associated with different chemical treatments. If compounds similar to the purines or pyrimidines were introduced into the cells, these "base analogues" would become incorporated into the multiplying viruses and they would lead to mistakes in pairing during or after incorporation. These mistakes would become fixed as permanent mutations. "The large mutagenic effect of our base analogues is probably due to mistakes in complementary base pairing, which accompany their incorporation into DNA. . . . In each pairing mistake, one purine is replaced by another purine, or one pyrimidine by another pyrimidine; i.e., in each single DNA chain, the number of purines, as well as the number of pyrimidines, is conserved at any stage of the mutagenic process. After sufficient DNA replications in a medium no longer containing the base analogue, mutant DNA will be present in which one base pair is replaced by another one. Because of the purine-pyrimidine conservation, the only possible base pair changes that can be induced by such base analogues are the 'transitions': $\begin{matrix} A \\ T \end{matrix} \longleftrightarrow \begin{matrix} G \\ C \end{matrix}$."[129]

In addition to the analogues, nitrous acid produced transitions; also, most of the high temperature mutations were transitions. Only 14 percent of spontaneous mutations were transitions. Any transition mutant could be reverted spontaneously or by any transition agent. Most of the spontaneous mutations and all of the mutations induced by acridine compounds, especially proflavine, Freese attributed to "transversions" or pairing errors resulting in the replacement of purines by pyrimidines (or the reverse). Thus AT could undergo transversion to TA or CG, and any other base pair could similarly be changed to two other types. Transition agents would not revert transversion mutations and transversion agents would not revert transition mutants. The spontaneous mutations were a mixed class, thus both induced transitions and induced transversions would revert spontaneously.*

Freese's general theory of mutagenesis made it seem plausible that individual mutant sites on a genetic map would be associated with changes of purines or pyrimidines. If the amino acid sequence of a gene-controlled protein were known, then there would be a good chance of using Ingram's techniques to isolate the mutant peptides with the amino acid replacements. By accumulating enough cases and using a computer, if necessary, to determine the restrictions on the numbers of substitutions at any one site, the code might be cracked. With this principle as a working model, Brenner and L. Barnett initiated studies on the genetic and chemical properties of the head protein of the bacteriophage T4. "It is well to specify the conditions which should be met by any system in which the relation between a gene and a protein is to be studied. On the genetic side, it should be possible to carry out a fine structure genetic analysis. This implies that mutants can be isolated, that recombination experiments can be performed, and that a selective technique is available for the efficient detection of low frequencies of recombination. Essentially this restricts the field to microorganisms, because only with these living systems can large enough populations be easily obtained and handled. The protein to be studied should comprise a relatively large fraction of the cell to make isolation easy and reasonable on a laboratory scale; the purification procedure should be as uncomplicated as possible, and the protein should have a reasonably small molecular weight so that the investigation of the amino acid sequence is not too laborious. Finally, it goes almost without saying,

---

* See Chapter 27, however, for a more generally accepted and alternative interpretation of Freese's "transversions."

that the gene to be studied should *directly control the structure* of the protein and not affect it indirectly through control of repressors or inducers."[46]

The rII region, which had a fine structure analysis that was well advanced, lacked a specific protein corresponding to it. Tail fibers did have a region—the host range (h) region—but the fibers only formed 1 to 2 percent of the total phage protein and the sub-unit was of massive molecular weight. Other components had similar difficulties. However, the head protein formed about 85 percent of the protein and had a sub-unit of 80,000 molecular weight, making it suitable for fingerprinting. Brenner and Barnett worked out a selective system to locate head protein mutants and to begin a fingerprint analysis of the mutants. This, however, was a slow process and other investigators hoped to use different proteins (but the same outlook) for such an analysis. In 1960, W. J. Dreyer reported on the progress that he and G. Streisinger were making on the small enzyme, lysozyme, whose molecular weight of 18,000, was ideal. "We were most anxious to see what sort of relationship exists between the genetic map of the various types of mutants in the lysozyme cistron of bacteriophage and a 'map' of the changes in protein structure. Recent development of chemical mutagens specific for particular purine or pyrimidine bases, together with new, relatively rapid methods for sequence analysis of proteins, and for fine structure genetic mapping, have made this sort of approach experimentally feasible. Thus we feel that within a few years results from work in progress in various laboratories will give an insight into the nature of the 'genetic code.' "[116]

The race was on. While geneticists were making a mental wager on which laboratory would be the first to solve the coding problem by the mutagen–fine structure–fingerprint route, an entirely unnoticed dark horse appeared on the scene, and with an entirely different approach, the coding problem was solved!

The solution came not from geneticists but from biochemists. M. W. Nirenberg and J. H. Matthaei at the National Institutes of Health were working on an *in vitro* system for protein synthesis. This system was successful. "A stable cell-free system has been obtained from *E. coli* which incorporates C^{14}-valine into protein at a rapid rate. It was shown that this apparent protein synthesis was energy-dependent, was stimulated by a mixture of L-amino acids, and was markedly inhibited by RNAase. . . ."[253] Further work with this system revealed an unexpected and completely novel discovery. "The present communication describes a novel characteristic of the system, that is, a requirement for template RNA, needed for amino acid incorporation even in the presence of soluble RNA and ribosomes. It will also be shown that the amino acid incorporation stimulated by the addition of template RNA has many properties expected of de novo protein synthesis. Naturally occurring RNA as well as a synthetic polynucleotide were active in this system. The synthetic polynucleotide appears to contain the code for the synthesis of a 'protein' containing only

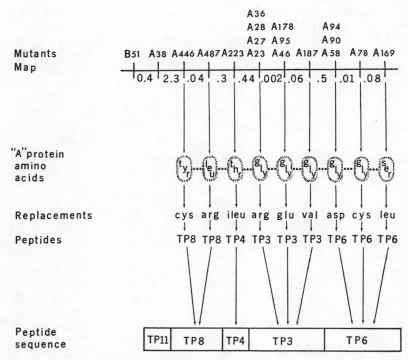

Figure 38. *Colinearity of Gene and Protein.* Yanofsky has localized the sites of the A region of tryptophan synthetase. These mutations cause amino acid replacements which result in a loss of enzyme activity. The tryptic peptides of the A protein show that closely linked sites are usually present in the same peptide. Furthermore, the sequence of tryptic peptides in the A protein is colinear with the sequence of sites in the genetic map. The colinearity is a consequence of the coding relation between the nucleotide sequence of the gene and the amino sequence of the protein specified by the gene.

one amino acid."[253] In their experiments, "the addition of 10mg of poly-uridylic acid per ml of reaction mixture resulted in a remarkable stimulation of C^{14}-L-phenylalanine incorporation. . . . Polyuridylic acid seemingly functions as a synthetic template or messenger RNA, and this stable, cell-free E. coli system may well synthesize any protein corresponding to meaningful information contained in added RNA."[253]

In the years since Nirenberg and Matthaei's discovery the coding problem has remained chiefly a chemical rather than a genetic study. The template RNA proposed by Crick turned out to be a smaller molecule than that larger portion of RNA found in microsomes or ribosomes. Today the template RNA is none other than the messenger RNA formulated by Jacob and Monod. The colinearity of the genetic map and the protein sequence has been established by several investigations, but in most detail by C. Yanofsky and his colleagues using the tryptophan synthetase regions of *E. coli*.[321] The analysis of the coding "dictionary" by synthetic ribonucleotides has been pursued by Nirenberg's group and by the laboratory of S. Ochoa, who was the first to achieve the synthesis of synthetic ribonucleotides *in vitro*. Crick's code was wrong. The present code is degenerate, and reading is accomplished by an initial and terminal "punctuation" along the sequence. The initial reading point is believed to be the operator. The nature of the signal for beginning and ending the reading of the sequence is not known.* There has been some speculation that only one chain of the duplex DNA is read and that this is associated with a circular organization of the transcribable unit (called the "transcripton" by S. Spiegelman).[163] This, however, has been challenged and the problem is still open. Of the sixty-four possible combinations that could code a triplet, at least forty-eight have been characterized, including all twenty-seven of the combinations of A, C, and U. The universality of the code still holds and presumably the degeneracy permits the variation in AT and GC content of different organisms. This "fourth stage" of the coding problem might be called the chemical stage. The precise molecular levels of mutation, transcription, and the energetics of the coding process are still tasks for the future. It is appropriate to state, however, that it is chiefly through the analysis of the coding problem that the study of the gene has acquired the interdisciplinary approach which was hoped for some forty-five years ago in Muller's prophetic view of the gene.

---

* In bacteriophage the triplet sequences (codons) UAG and UAA have been assigned such coding punctuation functions (mutations forming these codons are known as amber and ochre mutations). The natural sequences for such punctuation terminating and starting a gene message are still not known.

CHAPTER 26

Conclusions: Historical

Frequently, examples of discontinuity of progress in a science are dismissed as illustrations that the "times were not ready" for an understanding of the significance of a discovery. This may be true on occasion, but in genetics there are few instances in which it can be unequivocally asserted. The quest for the structure and function of the hereditary units is as old as the rediscovery of Mendelism. Bateson, Morgan, Muller, Demerec, Goldschmidt, Stadler, and Serebrovsky are only a few of the many geneticists who hoped to explore this problem prior to 1940. While biologists of the future will rightly date their knowledge of gene structure from the epochal discovery of genetic fine structure by Benzer, geneticists should recognize that the development of the problem is older than the prevailing solution.

Mendel's Patient Wait for Recognition

The most celebrated case of discontinuity in discovery is Mendel's long wait for his day of fame. Unfortunately, Mendel died long before the rediscovery of his contributions to heredity. Many geneticists have considered that the thirty-five-year delay in the recognition of Mendel's significant discoveries is a remarkable event; some attribute this delay to publication in an obscure journal, some to a lack of statistical knowledge among his contemporaries, some to the dazzle of Darwinism, others to the neglect and negative attitude of Nägeli, and still others to the inability of his contemporaries to appreciate his simple genius which was "ahead of his times." To some extent each of these interpretations has merit, but there is one aspect that I would like to offer on the basis of several cases men-

244

tioned in this book. I think it would have been possible for Mendelism to have had its start in 1865 and not 1900. The failure was due to Mendel himself. Unlike Bateson, Mendel did not debate with his critics. He did not stir up a storm with Nägeli. Unlike Bateson, Mendel did not widely publicize his results. Mendel, as far as is known, did not send a reprint of his work to Darwin although he was familiar with the importance of Darwin's contributions.

What would have happened if Bateson had sent a copy of his "Address to the Horticultural Society" to Weldon and Pearson before he had given it in public? Would Weldon's reactions have been different? Would he have welcomed Bateson's analysis as a monumental achievement? Certainly not; he would have informed Bateson of the numerous inconsistencies, the weight of evidence against its universality, the superiority of Galton's theories, and the inapplicability of Mendelism to these theories. In a very polite way he would have urged his friend not to make a fool of himself in public. After serious revision of the paper and some discussion of the potential contributions that this work might have to the laws of regression and familial inheritance, Bateson might, after all, have an article worthy of publication.

Similarly, what would have happened if Morgan, doubtful as he was about particulate models of heredity, had sent his first papers on *Drosophila* to Bateson? Would Bateson have been overwhelmed by the implications of "associative inheritance" or "chiasmatypie" as a mechanism for reduplication? Again, we would expect Bateson to have argued strongly against its publication without extensive modification. He might not have rejected it outright, but he certainly would have raised numerous objections which would have overwhelmed Morgan if his personality lacked self assurance. For similar reasons it is difficult to see why Nägeli should be blamed at all for criticizing Mendel's experiments. He did what every geneticist is expected to do when he receives a manuscript from *Genetics* or any other professional journal. A referee would merely be a rubber stamp if he did not assume the role of devil's advocate in judging his colleague's paper. If a scientist is too modest to assert the significance of his results, if he shrinks from the possibility of scientific debate, then he alone is to "blame" for the failure of his discovery to be made known. It is as unfair to blame Nägeli as it would be to blame Weldon or Bateson in the hypothetical examples cited.

This illustrates the role of the scientist in discovery. The discovery may be intrinsically great, but its greatness is often made apparent by its discoverer or its rediscoverers and admirers. Personality is as important as factual content in the development of a concept, particularly so in the establishment of the concepts that dazzle and dominate the scientific community. The interpretation of data of the stature of Mendel's requires missionary zeal to establish it because of its novelty and lack of predecessors. Bateson had just that in 1902 when he took on all of his scientific community and forced Mendelism upon them. Morgan early in 1912 recognized the worth of his own ideas and urged his students to collaborate with him in the writing of *The Mechanism of Mendelian Inheritance*. Muller pulled no punches in fighting for the significance of the individual gene and of mutation as the tool for its analysis. This concern for one's discoveries is not mere egoism. It is a commitment made by the scientist to the worth of his studies. Mendel lacked the abrasive aggressiveness of Bateson and Muller. He was too humble to fight for his cause, to put his scientific ego in the forefront of his activities. If this is so, then Mendel was the wrong man to discover Mendelism!

Discontinuity in Discovery

There are several other instances of discontinuity in discovery that I have illustrated in this book. One, in particular, deserves discussion. In 1921, at Toronto, a symposium on variation was held under the auspices of the American Society of Naturalists. Three major events occurred at the meeting. For the majority of the geneticists in the audience it was a cause for rejoicing—Bateson acknowledged the relation between chromosomes and the hereditary factors. For the public at large it was the beginning of a scandal that would lead three years later to a showdown in Dayton, Tennessee. Bateson's address "Evolutionary Faith and Modern Doubts" had pointed out that the new gene theory had removed the old fluctuating Darwinism of Galton's school as well as de Vries's mutation theory as contenders for the mechanism of speciation. He urged more attention by geneticists to the problems of evolution and particularly to its mechanism. This talk was misinterpreted by reporters whose headlines ("Darwin Downed") excited William Jennings Bryan into advocating the abolition of evolutionary doctrines from the public schools and universities of the United States. The third event, more momentous than either of these two, was heard by the geneticists and scientists at Muller's discussion of "Variation due to change in the individual gene." Muller had drawn the striking parallel between bacteriophage (d'Herelle bodies) and genes. He cited their potential for genetic research and their eventual utility for a common approach to the gene problem by physicists, chemists, microbiologists, and geneticists. While Osborn scoffed,* others heard. Yet, no one took up the lead. Why? Again the fault lies with the person who proposed a novel concept. It is unlikely that the students of Twort and

---

* According to Muller, Henry Fairfield Osborn thought Muller was providing a playful hoax on the audience. After the speech was over Osborn commented, "It's a good thing that you have a sense of humor, Dr. Muller." Others in the audience were impressed by the vigor of Muller's presentation. I asked Dean Rollo Earl of Queen's University (Ontario) if he had heard this speech in Toronto. "Yes," he replied, "I can still hear Muller's words, 'Let us hope so!' "

d'Herelle were in attendance at a meeting addressed primarily to students of evolution and genetics, because they would have had no reason at that time to associate chromosomal inheritance with viruses. But even if they were passively exposed to these prophetic remarks, they probably would have ignored them unless Muller himself made a strong effort to urge the participation of virologists in this search. Bateson would have found few converts to Mendelism had he not gone to the breeders themselves. Discontinuity, like continuity, in science, is subject to the strength of interest of its contributors.

The strength of ego alone is insufficient to establish a concept or discovery when the data are scant or vague or if the model itself is unproductive. Castle fought for his convictions as hard as Bateson did for his, but he failed, twice, to win over the rest of the scientific community. In both instances he was no more right than wrong before the critical experiments were performed disproving his hypotheses. Yet prior to these experiments he was losing the battles. It was not that Morgan, Johannsen, and East were more eloquent; Castle did not lack in literary skill or sharpness or wit. But "residual" heredity made possible far more predictions than did allelic contamination. It readily accounted for the blended inheritance of size and weight while allelic contamination made only a half-hearted attempt at explaining this phenomenon. Similarly, Goldschmidt expressed a forceful personality in his articles; he was not afraid to criticize prevailing theories; he was far more eloquent than most geneticists in his use of English even though it was his second language. But most of Goldschmidt's contributions to the gene concept were based on his adversaries' data rather than on discoveries of his own researches. His scant data never matched the scope of Muller's prolific discoveries from 1927 to 1940. His theories suffered, also, because they were not positive assertions but criticisms—they served as a dam constructed to hold back the flood of implications pouring out of Muller's interpretations of the individual gene.

In one man an exceptional illustration can be made of the influence of fact and personality on the success of a theory. Demerec was unsuccessful in establishing the genomere hypothesis in 1927, but he was successful in advocating and co-discovering fine structure of the gene in 1955. Fortunately, Demerec's enthusiasm for the study of the gene was not discouraged to the point of withdrawal when he was unable to find support for his genomere hypothesis in *Delphinium* or when radiation proved unsuccessful in providing more mutable genes in *Drosophila virilis*. Demerec succeeded in 1955 because the techniques available for the analysis of his model were more direct, and his model was based on the linearity of the genetic material. Yet the linearity of the genetic material was suggested by Muller in 1918 on cytological grounds, and it was this linearity which Demerec rejected in the 1920's. Through variegated pericarp investigations, Demerec's training as a maize geneticist provided him with the genomere model. The maize model of variegation, in turn, traces its origins to Correns's pre-chromosomal speculations on "sick genes." Thus the tools of Drosophila research were applied to a model which lacked or rejected the very information in Drosophila which would make the geno-mere model unlikely. On the other hand, Demerec was using a model of the genetic material in 1955 that was cytologically and biochemically expected to be linear. His discovery of intraallelic recombination was consistent with the prevailing assumptions and knowledge of bacterial genetics.

Research Eclipse

A second feature of the process of development of a concept is a phenomenon that might be called research eclipse. There are several examples of productive methods and ideas which reached an ascendance and then faded into obscurity only to be resurrected many years later, some-times in the same organism, sometimes by the same investigators, and sometimes with entirely novel designations with new techniques and organisms used for investigation. Position effect was dominant in the 1930's; its revelation of the necessity for fixed orders of genes in linear series was unexpected by most, if not all, geneticists. Its potential for attacking the concept of the individual gene was realized by Muller him-self, who first suggested this as an alternative mechanism for radiation-induced mutation. The striking physiological effects of variegation, their relation to heterochromatin, and their modification by added hetero-chromatin and alteration of temperature all afforded immense opportunity for further research. Yet few novel concepts emerged as a result of these discoveries. Interest in position effect fell into obscurity during the 1940's and early 1950's. The rise of the operon model in 1960, however, has given this phenomenon a renaissance. Many of the features of the operon model could have been proposed in the 1930's: position effect was the basis for most dominant mutations; it resulted in repressible rather than permanent changes; and the gene affected by position effect could be extracted with normal function from the new alignment. The Drosophila group had turned genetics inward, using phenotype and the gene concept as the tools for studying the chromosomal mechanics of hereditary transmission, organization, and mutation. Differentiation, cellular specificity, organ for-mation and its genetic control—the dreams of Morgan which had failed to materialize—were forgotten by geneticists and remained in an eclipse period of their own.

A second example of research eclipse is the "left-right test." It is surprising that no locus other than scute has been explored by this tech-nique. It is theoretically possible to use it in any system which yields viable structural changes. Maize research could readily use this technique, yet no one has ever attempted it, at least not on a scale provided by the analysis of the tip of the X chromosome. Part of the difficulty with the "left-right" test is its exceptional sophistication. It requires a good deal of genetic analysis to establish the cytological break points of rearrangements and it requires ingenuity to render material co-isogenic to eliminate modifier effects on the phenotypes of the mutants and their "left-right" combinations. Despite these limitations, the significance of the problems of chromosomal organization is worthy enough to recommend its resurrection in genetic research.

Still another example is step-allelism, almost unbelievable in its first bold form in 1929. Yet precisely that model was employed twenty-five years later in microbial research under the name "complementation mapping." It is likely that the microbial phenomenon was rediscovered rather than stimulated by the Russian school of geneticists. This area of research would probably have been revived in the U.S.S.R. if the Lysenko controversy and the war had not disrupted the participants and their research programs, but even without Lysenkoism, it would have fallen into eclipse until the late 1940's or early 1950's because of the criticisms it received from other *Drosophila* geneticists.

The habits of a scientist in the course of his career reveal a tendency to replace, rather than retract, his model or erroneous concepts and interpretations. Thus Morgan, between the years 1900 and 1915, changed his views several times on the roles of chromosomes and factors in heredity. He never formally retracted these earlier ideas; he allowed them to wither into oblivion by neglecting to cite them or by modifying them in subsequent papers. Castle, on the other hand, retracted his errors not only because he felt it was the decent thing to do but because he had no choice but to concede. Castle did not present a variety of possibilities from which his readers could choose; he forcefully singled out one model alone as the interpretation of his data. His commitment to his theory of allelic contamination was so strong, and his own data contradicting this hypothesis were so unquestionably convincing to him, that he had to retract. On the other hand Goldschmidt never proclaimed his earlier "dynamic" model of quantitative components of the gene to be wrong. He merely rejected the entire concept of the gene. He never acknowledged that the Bar case, as he interpreted it several times for his different models of the gene, had been previously misinterpreted by him. Goldschmidt was not forced into retraction because the status of gene mutation and gene structure was still speculative. Retraction is only necessary when an interpretation has achieved such stature that the scientific community accepts its validity.

Research Impasse

If research eclipse represents productive techniques or concepts that remain untouched for years, then *research impasse* represents the failure of competitive theories or models to achieve resolution in the absence of critical experiments. The two outstanding cases are Muller's controversies with Goldschmidt and Stadler. The criteria, operational or otherwise, have not yet been developed in Drosophila. Presumably successful mutagenesis with base analogues, occurring only during replicative stages in early meiosis or early cleavage, would be highly suggestive. Die-hard skeptics, however, would probably await sequence analysis of nucleotides of the affected gene before accepting intragenic mutagenesis of a chemical nature. Neither Goldschmidt nor Stadler considered the left-right test as evidence for discontinuity in the organization of the genetic material. It would have been a valuable contribution to the evaluation of this technique if either of these critics had stated their objections. The future use of the left-right test at other loci would again constitute highly suggestive evidence for genetic discontinuity. The true skeptic would probably hold out for more direct, biochemical evidence.

Where the issue in research impasse is almost exclusively a genetic one, such as Bateson's adherence to reduplication rather than to crossing over, the controversy is resolved to all contending schools by the failure of one model to predict as successfully as the other model. Even in this instance, however, individual skeptics remained (without a following) until the cytological techniques of altered chromosomes (as used by Stern and by Creighton and McClintock) and salivary chromosome analysis demonstrated beyond doubt the general validity of the linear localization of the genes along the chromosome. In Castle's two controversies with the Drosophila group, the issues were genetic and resolved without cytological or biochemical techniques for independent confirmation. Both of Castle's models were withdrawn as soon as he was forced, by his own experiments, to contradict the predictions of his genetic models. These unexpected results, in both cases, were compatible with the models used by the Drosophila group: modifying factors rather than allelic contamination in the one case, and the linearity of the genes with crossing over and interference rather than with a three-dimensional construction in the second case.

Where two contending views exist, much dissension can be avoided if the problem can be dissociated into the levels of experimental analysis needed for its resolution. Is the problem solely genetic? Is a cytological or a biochemical problem actually involved rather a genetic one? The controversy between fine structure and pseudoallelism in Drosophila is illustrative of this difficulty. A multiple allelic system showing one phenotype (e.g., eye color) would not be resolved by the finding of a few separable sites. If large numbers of sites were obtained, approaching twenty or more, it would be inconceivable that a single region would be composed of so many separate "genes." In no case has this been done with exhaustive care in Drosophila. On the other hand, the controversy could be resolved with a small number of crossovers if a multiple allelic system such as Notch, dumpy, or bithorax were shown to have a few sites for one of the pseudo-allelic regions exhibiting a specific phenotype. If several alleles of one pseudoallele, in any of these complex loci, were separable from one another, this would force recognition of fine structure or require a second level of sub-units as an ad hoc modification of the theory of pseudoalleles.

In the history of this controversy, however, pseudoallelism fell into disfavor with the ascendance of the more attractive molecular model of fine structure; it has been revived because the operon model is similar to the pseudoallelic model. The critical demonstration of fine structure in Drosophila awaits a genetic solution which requires only patience and time. The techniques and systems are both available. In bacteria the resolution was immediate because the number of sites in a region can be used to define the structure as one of pseudoallelism or fine structure. Nevertheless, the linked systems of genes discovered by Demerec for tryptophan synthesis and by Benzer for the two rII cistrons suggested that an intergenic relation as well as an intragenic relation existed among these genes. The extension of this observation to coordinated function in the operon model, in turn, forced the reconsideration of pseudoallelism in Drosophila.

The Fallacy of Occam's Razor

One of the philosophical principles frequently invoked during research impasse is that of economy of hypotheses. Occam's razor, however, is not usually a valid instrument for the resolution of contending models. It was used unsuccessfully by Castle against the "ad hoc" theories of modifying factors and interference. It was used implicitly by Bateson in justifying "presence and absence" over the dual system of chemical and physical alterations of the gene advocated by the Drosophila group. It has been explicitly stated by Goldschmidt in his attack on the gene concept which "had to be saved" by position effect and other "ad hoc" constructions. It was an explicit criterion used by Stadler in advocating breakage as the exclusive mechanism of radiation mutagenesis in plants and animals. It was also implicit in the attempts by Pontecorvo between 1954 and 1960 to interpret all multiple allelic series as "cistrons."

Why then, with so many examples of its failure, is Occam's razor so popular among scientists? I believe that scientists who use this principle to oppose a contending hypothesis are unaware that their own models are incomplete; their use of Occam's razor is thus invalid. Its usage also implies an aesthetic judgment based on order, symmetry, and simplicity. It may be useful when applied to one's own model, alone, but I do not believe it has any scientific merit in the resolution of a controversy because it avoids the need for an exploration of the different levels of experimental analysis; it evades a critical study of the predictability of different model systems, and it evades the need for the comparative examination of a genetic principle in several organisms and in more than one experimental design.

Scientific Progress: Human Limitations

Scientific progress is a consequence of intense human activity. In genetics, the direction of research, the prevailing outlook, the choice of organism, the reading habits of the scientist, the preparation of textbooks, and the content of college courses are dictated by the personalities of the participants. It is this human quality which gives genetics its greatest strength as well as its chief weakness. To the proselytizing efforts of Bateson we owe the initial intensity of interest in genetics. We may pay homage to Correns, de Vries, and Tschermak for rediscovering Mendelism, but only de Vries had the stature and personality which commanded the attention of the scientific community. However, he lacked interest in the phenomenon he had rediscovered because he was convinced that his greater fame rested on the replacement of Darwinism with his hypothesis of de novo speciation by macro-mutations. If it were not for Bateson the direction of research would have been turned toward evolution theory and not toward heredity itself. Eventually the weaknesses of de Vries's "mutation theory" would have been recognized, but a delay in the development of the gene concept would probably have occurred.

To the fortunate juxtaposition of Morgan and Wilson in the same department of zoology we owe the stimulation which led to the Drosophila group. Our tributes are usually directed toward research systems, but a teaching climate in this case brought Morgan, Wilson, and their students into a comparative examination of the chromosome theory which would have been difficult to achieve had either of these teachers not been there. To Morgan himself we owe appreciation for his flexibility, his readiness to recognize and incorporate novel concepts, and his voluminous efforts to popularize the discoveries and direct implications of the "gene theory," which he and his students had developed.

Muller's contributions dominate the 1920's and 1930's. The variety of genetic tools which he developed elevated Drosophila to an almost exclusive role in the study of the gene concept. It is impossible to read Muller's articles during these two decades without feeling overwhelmed by the quality of genius which characterizes the analytical and projective treatment of his discoveries and experiments. Yet this very quality of genius had a seductive effect on scientific progress. It drew talented investigators into Drosophila genetics; it focused the attention of geneticists on radiation as the exclusive mutagenic source; it eclipsed interest in developing similar intense explorations of the gene concept in other organisms; and it eclipsed interest in physiological and developmental genetics. I pointed out before that Muller could have initiated an interest in bacteriophage genetics in 1922 if he had desired to expend the effort to do so. It is impossible, however, for one man to throw his energies into more than a few projects. For Bateson there was only one: Mendelism. For de Vries there was only one: Oenothera. For Morgan there were two: Drosophila genetics and marine biology. For Muller there was "mutation and the gene." Muller chose to remain with Drosophila because the tools were developed, the ideas were plentiful, the problems were well defined, and the techniques were free from technical error.

The tragedy of genius is that it is associated with its possessor. His choices among the many facets of his genius eclipse the remaining facets; it is more a tribute to chance than to design if the secondary facets make an impression on the minds of his followers and admirers. Genius is simultaneously inspiring and inhibiting; it may force the bright and the brilliant out of competition altogether; it may attract others, equally brilliant, into a communion of scientific interest. When genius is coupled with the urge to communicate and proselytize, it has its most dramatic effect. To Muller we owe the transformation of the gene concept from an inchoate term to the physical model of a linear ribbon with dimensions, boundaries, properties, organization, and stability: each of these features being characterized by his experiments in the two decades between the World Wars.

The Necessity for "Straw-Man" Models

Beginning with Goldschmidt's attacks on the gene about 1917, the gene has frequently been depicted as a spherical "bead" on a gene-string. This "classical gene concept" is a fiction. It is based on the analogies and illustrations used in the development and popularization of the "factorial hypothesis." With the maturity of the experimental investigations in Drosophila between 1914 and 1918, this "classical" model ceased to exist. Multiple allelism, mosaicism, detailed mapping, and the cytological discoveries of chromosomal continuity throughout the mitotic cycle were all major contributions to the definition of the gene with a structure more complex than a factorial bead. A careful reading of the literature of the 1920's would not fail to reveal how much controversy and theory existed on the details of the gene concept. The repeated use, however, of this "straw-man" concept of the classical gene persists. In the hope of dispelling this fiction from the history of genetics, I have tried to point out the variety of models proposed for the gene. Whether Castle, Eyster, Demerec, Goldschmidt, East, Correns, or Muller is cited in the development of the gene concept before the introduction of radiation mutagenesis, the result is the same: each discussed uniquely different models or properties of the gene which were opposed to the "straw-man" model of the classical gene.

In the 1930's the variety of models increased not only through position effect analysis but from the studies of step-allelism, minute intergenic and intragenic rearrangement, and the left-right test. The target theory provided a "sensitive volume" two to three orders of magnitude smaller than the volume estimated by the cytogenetic techniques of Muller and Prokofeyeva. It is difficult to find a single contributor to the gene concept in the 1930's who invoked, explicitly, a "bead-on-the-string" model of the gene, except in the guise of a "straw man."

There is a tendency, in the process of rediscovery or independent discovery, to counterpoise the new hypothesis or model with the weakest possible model of the prevailing view. The sharpness of contrast between the two models permits a fresher, freer outlook for the new model. Thus the repudiation of Bateson by the Drosophila group permitted a total transformation from *Mendelism* to the *Mechanism of Mendelian Inheritance*. The failure to give proper credit to Bateson for foreseeing many of the aspects of the "new" genetics resulted from Bateson's failure to popularize these predictions by repetition and experimentation as well as from the psychological tendency to discard the totality of an opponent's contributions in the presence of those dominating errors which constitute his prevailing views at the time of a controversy. Similarly, fifty years later, it is not surprising to find some of the proponents of linear molecular models of the gene frequently invoking a "straw-man" classical gene as the pattern of enslavement from which they escaped!

The "straw-man" habit of strengthening new concepts is useful for research, but it falsifies history in the interests of establishing new patterns of research. I believe this scientific habit is worth maintaining; the scientist owes more to the present and the future than to the past. It is the obligation of the historian of science, not the scientist, to reconstruct the development of concepts and to restore the fallen idols of previous generations to their original eminence.

The Ambiguous Role of Philosophy in Genetics

The magnitude of the contribution of a geneticist to the gene concept is not dependent on the implied or explicit philosophy of science that he adopts. Among the explicit philosophies advocated in the past are mechanism, materialism, dialectical materialism, and operationalism. Vestiges of idealistic philosophies are present in the inaccurate use of Occam's razor, in holistic models of gene actions, and in the purely conceptual, nonphysical, factorial models of the gene. To label any particular model as belonging to a philosophical school is risky because frequently the philosophical aspects of a model are not of particular concern to its proponent. Thus Morgan considered himself a mechanist, but his earlier interpretations of heredity reflected antagonism to such mechanistic interpretations as Sutton's and Wilson's chromosome theory of inheritance. Muller considered himself a dialectical materialist until the later 1930's when this philosophy itself became the basis of an attack on the gene concept and the chromosomal basis of heredity. He never declared himself to be an operationalist, as Stadler declared himself to be, yet his criteria for distinguishing the gene on the basis of mutation, replication, recombination, function, and breakage are virtually indistinguishable from those advocated by Stadler, who explicitly attributed his criteria to operationalism.

During the late 1940's, after the Lysenko affair had become a political issue, Muller repudiated dialectical materialism because it extended beyond the assumed right of scientists to decide among themselves, on the basis of their competitive models, which views would prevail. This "right" however, is not inherent in the public attitude. There was an apocryphal story I had heard as a graduate student about a teacher who, in 1948, read to his class a "letter" supposedly from President Truman, who officially condemned Lysenkoism as a perversion of scientific truth and officially endorsed the chromosome theory as correct. The genetics class cheered this "endorsement" and received an unexpected lesson in the philosophy of science for their enthusiasm for this false report.*

While official policy may restrict the development of a science as it did during the Lysenko affair and during the years of Hitler's racist "Third Reich," it offers little impairment to those sciences not under political attack. Great discoveries have been made under absolute monarchies, totalitarian dictatorships, and democracies. Furthermore, intimidation of scientific concepts cannot exclusively be attributed to totalitarian systems. In the United States, the wave of anti-Darwinism that was launched after Bateson's address on "Evolutionary Faith and Modern Doubts" has never completely disappeared. It is still against the law to teach evolutionary theory in the public schools of Tennessee although the actual enforcement of the law has been largely ignored since the Scopes trial. Bateson cited "secret" societies in some schools in Washington (prior to 1920) which discussed evolution theory in defiance of official college decrees to keep it out of the biology curriculum.

It is sometimes useful to explore the philosophical basis of a scientist's concepts and approaches, but there are also disadvantages for the scientist when he attributes his own concepts, or the concepts of an opponent, to philosophical schools. To do so impugns the individuality and, frequently,

---

*Lysenkoism has not fared well in the U.S.S.R. since 1948. After Stalin's death, Lysenkoism was downgraded, but over the past 15 years it has had periodic revivals. Its status today is at its lowest. There are two new journals of genetics in the U.S.S.R. which publish articles comparable to those found in the major journals of genetics used as references in this book. A recognition of the contributions to genetics of N. P. Dubinin was made in 1966 with the award of the highest Soviet scientific honor, the Lenin Prize. Also, a yearly prize in genetics has been established in honor of N. I. Vavilov, the geneticist who had befriended Muller during his stay in the U.S.S.R. Vavilov, who had been arrested in 1939 as a "British spy," died during World War II. His famous contributions to the genetics and evolution of agricultural plants are being republished by the U.S.S.R.

the sincerity of another scientist's work. It also "dates" the scientist, himself, if he attaches a philosophical label to his theory; when newer philosophies emerge, the older ones tend to detract from, or eclipse altogether, the merit of the scientific contribution itself.

The Comparative Genetic Outlook

In surveying the history of the gene concept, I have found one point of view, frequently alluded to, that I should like to advocate vigorously. This is the necessity for a "comparative genetic outlook." This outlook is based on the premise that research with a single organism or a single technique restricts the possible universality of new concepts The differences between maize and Drosophila in response to x-rays is one example of how two views of mutagenesis could remain in contention so long as each protagonist remained indifferent to the differences manifested by another organism. Fine structure in viruses and bacteria may be demonstrably different from pseudoallelism in Drosophila where definitive fine structure is lacking. Many geneticists have assumed that all multiple allelic series in Drosophila are "cistrons," despite Benzer's preference that complex loci not be called cistrons unless they meet the rigid operational criteria proposed for them.

It is sometimes profitable to read outside of one's own research area because the implications of one system, applied to another, may produce unexpected results. I wish to use an example from my own work as an illustration. In 1928 Muller advocated that most, if not all, x-ray induced mutations were fractional. The degree of strandedness of the chromosome in gametes, however, was limited to two because no mosaic sex-linked lethals could be detected after the F_3 generation. When chemical mutagenesis was demonstrated by Auerbach and Robson, during World War II, the production of mosaic visible mutations in Drosophila was its most remarkable feature. No serious effort, however, was made to detect sex-linked lethals in the F_3 or later generations at that time. This was probably a purposeful neglect based on Muller's findings in 1928. In the mid-1950's microbial mutagenesis with base analogues indicated mosaicism or sectoring would be the rule for replicative errors. My own work on the dumpy locus, using x-rays, indicated a fairly high degree of mosaicism, about twenty-five percent. With the collaboration of I. I. Oster, I sought an agent which would produce mutations in Drosophila with as few gross breakage events as possible. We used a monofunctional quinacrine mustard which answered that need and produced an abundance of mosaics at the dumpy locus, so high, in fact, that it was difficult to believe that any complete or "non-fractional" mutations were induced at all. This finding, supporting microbial models of mutagenesis, led to our deliberate search for a far more extensive degree of mosaicism among sex-linked lethals. The extensive analysis of this sex-linked lethal mosaicism, especially by J. L.

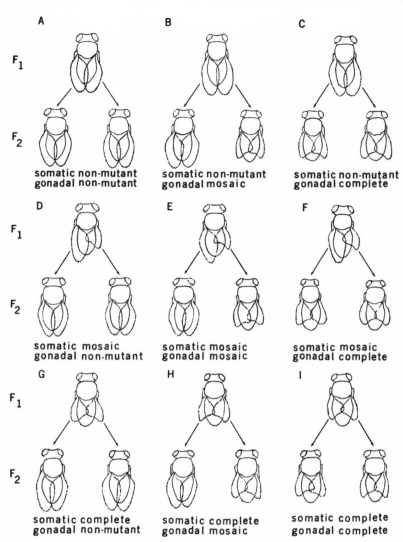

A

F₁

F₂

somatic non-mutant
gonadal non-mutant

B

F₁

F₂

somatic non-mutant
gonadal mosaic

C

F₁

F₂

somatic non-mutant
gonadal complete

D

F₁

F₂

somatic mosaic
gonadal non-mutant

E

F₁

F₂

somatic mosaic
gonadal mosaic

F

F₁

F₂

somatic mosaic
gonadal complete

G

F₁

F₂

somatic complete
gonadal non-mutant

H

F₁

F₂

somatic complete
gonadal mosaic

I

F₁

F₂

somatic complete
gonadal complete

Figure 39. Gonadal and Somatic Mosaicism in Drosophila. If mutations arise in a mature sperm, one or both strands of the genic DNA may be affected. The distribution of mutant and non-mutant DNA strands to the nuclei of the cleavage divisions may result in any of the nine patterns illustrated. Only patterns C, F, and I are detected in a routine F_2 sex-linked lethal test. Mosaic sex-linked lethals which are detected in the F_3 may be obtained from patterns B, E, and H. A specific visible test using the mutant dumpy will detect these six patterns as well as the non-transmitted patterns D and G. Note that neither of the sex-linked lethal tests can detect patterns D and G. In a microbial colony which manifests sectored mutations, all mutations will be detected as sectors of varying size. The metazoan is a three-dimensional differentiated colony whose "sectors" of mutant tissue are distributed non-randomly to gonadal and somatic tissues.

257

Southin, and its comparison to the various classes of visible mutagenesis at the dumpy locus has made it possible to look upon a mosaic mutation in Drosophila as a three-dimensional sectored colony of cells. In this model an apparent phenotypic dissimilarity exists between a clone of cells in a microbial colony (essentially two-dimensional) and a clone of differentiated and highly organized cells in a fly. This difference may be resolved if the various aspects of determinate and indeterminate embryonic cleavage, morphogenetic movements, and other aspects of the fly's embryology are taken into account. This embryological model represents an attempt to use the "comparative outlook" to generate a model.

The "comparative outlook" does more than generate models. It permits the geneticist to seek ways out of the genetic impasse that often results from the presence of contending models in one system. It has its chief value in helping to prevent such research eclipse as the delay in rediscovering Mendelism, the delay in introducing viral genetics, and the delay in applying Garrod's "inborn errors of metabolism" to the gene concept. A mastery of the techniques for more than one organism is seldom achieved by an individual, but the exposure to the concepts generated from experiments with different organisms can give a geneticist a more enriched way of looking at his own material and its implications.

CHAPTER 27

Conclusions: A Comparative View of the Gene Concept

The gene has been considered to be an undefined unit, a unit-character, a unit factor, a factor, an abstract point on a recombination map, a three-dimensional segment of an anaphase chromosome, a linear segment of an interphase chromosome, a sac of genomeres, a series of linear sub-genes, a spherical unit defined by target theory, a dynamic functional quantity of one specific unit, a pseudoallele, a specific chromosome segment subject to position effect, a rearrangement within a continuous chromosome molecule, a cistron within which fine structure can be demonstrated, and a linear segment of nucleic acid specifying a structural or regulatory product. Are these concepts identical? Do they reflect a historical refinement of technique making the most recent definition of the gene the most valid appraisal of the concept? Can the gene concept differ in different organisms? If there is only one universal "composite gene," the solutions to the problems of gene chemistry, physical organization, regulatory function, continuity, strandedness, mutation, and coding must be drawn, at present, chiefly from Drosophila, maize, bacteria, and bacteriophage, as well as from inference and synthetic systems. Not all aspects can be drawn from one organism. For some of these problems, the findings from different organisms are contradictory; for others, the agreements may be analogous rather than a reflection of identical genetic organization.

If we assess the gene concept with these problems in mind, we can make an appreciation and evaluation of what has been established and what needs to be done. In agreement with the work of Muller since 1914, we find that the gene does control the specificity of all experimentally

259

discernible features of the organism—its visible features as well as its physiologically more significant components, including all the major molecular participants in metabolism. The contemporary view agrees that mutation is the chief tool for the genetic analysis of the gene. The basic techniques for calculating the number of genes, the size and shape of genes, and the finding that the gene is a linear molecule are also part of the Mullerian heritage. Additionally, the belief that replication is a result of a unique structural feature of the gene, independent of mutational change ("convariant reproduction"), has been confirmed. The bipartite structure of the haploid material, which was inferred in 1928 from x-ray induced mosaic mutations, has found its ultimate explanation in the chemical basis of heredity. Whether genes are considered "leaky" or hypomorphic, "non-leaky" or amorphic, this physiological interpretation traces its origin to studies of dosage effects by Muller and by Stern. That genetic material is physically or functionally discontinuous is a moot point, but the discontinuity is as real for Drosophila as it is for the inferred reading points which begin or terminate the gene transcriptions in bacteria. All of these various aspects, derived from Muller's work in Drosophila over a twenty-year span find agreement with features obtained in more recent years with microbial systems.

The contemporary models of the gene agree with Stadler's finding of functional complexity in a single genetic or allelic region. Also, bacteriophage, like maize, respond to x-rays as if their sensitivity for breakage swamped out any detectable gene mutations. The contemporary models also extend the findings of Dubinin and Serebrovsky on step-allelism, suggesting that complementation is found in all organisms, but not in all loci, and that this complementation can be mapped. In both cases, however, the promise of complementation topography for a structure of the gene or of the product of a gene has been less successful than its original expectation. The contemporary models also agree with Goldschmidt's contention that recombination may be intragenic and that there are hierarchies of sub-units depending on the criteria used for their definition. Lewis's work on pseudo-allelism has illustrated a feature associated with current microbial models of the gene: a complexity greater than that of the cistron, associated with sequences of genes corresponding, in part, to the biochemical pathways they control. There are features of Lewis's model strikingly similar to those of the operon. The contemporary models are also in agreement with Beadle's suggestions that the individual gene provides the specificity for a single primary characteristic which is, moreover, usually associated with an enzyme.

In the graveyard of genetic concepts are the unit character, allelic contamination, notational genes, overlapping genes, mutation exclusively by presence and absence, and the concept of the chromosome as a single gene or pattern. In various degrees of moribund condition are the hypotheses that all mutation is due to breakage and rearrangement, that colinearity exists between complementation maps and recombination maps, that the sensitive volume calculated by target theory corresponds to a gene, that cytoplasmic inheritance involves plasmagenes, and that all multiple allelic series are cistrons. In the realm of legend, I would place the "classical" bead-on-a-string gene.

Among the current features of the gene concept, some have not been subject to critical experiment and even fewer have been evaluated by a comparative analysis of their similarities and contradictions. In this final chapter I shall raise some of these questions and give my own evaluations with the hope that they will stimulate reflection and interest in the gene concept. For convenience the problems are distributed among five questions: Is the cistron too rigorously defined for analysis of the functional gene? Can position effect be interpreted by the operon? Is the physical organization of the chromosome linear or circular, continuous or discontinuous, single-stranded or multistranded? Is point mutation a definable concept? Are there major differences, other than specificity, which distinguish the genomes of microbial systems from those of higher organisms?

The Cistron as a Functional Unit

In its ten-year history, the cistron has stimulated a great deal of admiration and considerable, but not universal, acceptance. Its major contribution, the demonstration of genetic fine structure within the gene, has provided a genetic basis for molecular models of gene structure and function. It should be recalled, however, that the concept of fine structure was proposed and demonstrated in the rII region of phage two years *prior* to the formulation of the cistron. The cistron as a functional unit was developed with a class of mutations which were non-reverting and non-leaky. In *Drosophila* and similar multicellular systems such mutations would correspond to non-revertible amorphs or minute (intragenic) deletions. If a similar, rigidly defined class of mutations was used for multiple allelic series in *Drosophila*, few if any of the pseudoallelic systems known would have been discovered. The hypomorphic mutants used for pseudoallelism in *Drosophila* correspond to the leaky mutants which do occur but which were not used for analysis of the rII region. Such mutants should be tested in the rII region to obtain their sites on a fine-structure map. Are the two systems co-extensive? Are there discontinuities in the functional gene with the cistron forming only a part of its total length? Are all sites capable of forming amorphs and hypomorphs?

The cistron defined by the use of non-leaky mutants corresponds to a non-complementary set of alleles. The cistron concept would be incorrectly applied to complementing genes even if the mutants used in the study were non-revertible and amorphic. If the number of complementing genes in an organism is high, the cistrons (as defined by Benzer's standards) might represent only a minority of the genes in the organism.

In Drosophila, maize, mice, and other higher organisms, the cistron could be applied by analogy, not by definition, to complex loci or pseudo-allelic series. This would hold true even where the phenotype of the alleles is essentially identical. Allelic series such as vermilion and rosy eye color affect one character, but their fine-structure map is limited to a few sites. More complex regions, such as Notch, bithorax, and dumpy, are as readily interpreted by operons as they are by pseudoallelic nests of genes. For these reasons the term "functional gene" or "gene" alone is more accurate when critical demonstrations of a cistron are lacking. The cistron is a clarifying concept, but where critical analysis is lacking it should not be applied. If extensive complementation exists in a region with detailed fine structure, then the usefulness of the concept can be questioned. There are at present no genes in Drosophila which can be called cistrons, not because they may not exist there, but because an effort to rigorously demonstrate them has not been made.

One implication of genetic fine structure is the definition of a lowest limit of resolution by recombination. If this were not the case the meaning of a fine-structure map would be questionable. The discovery of "ultrafine structure" in the rII region by I. Tessman raises this problem. According to this finding, there is no apparent lower limit among certain members of "hot spots" in a fine-structure map. The fineness of these "recombinations" is about 100 times smaller than the recon defined by Benzer. If true, the molecular correspondence of the recon to one or two nucleotide pairs would no longer exist. This suggests that recombination is not the mechanism for the wild-type plaques obtained in ultrafine structure. One possibility is that these are reverse mutations, increased in frequency by the occasional congruence of the mutant regions at the time of replication. If pairing errors were increased at this site for either strand because of the disturbing presence of the homologous mutant site, then reverse mutations might occur more frequently than would be expected spontaneously for either mutant alone. This model is analogous to the "variable force hypothesis" proposed by Goldschmidt for Morgan's recombination data. Like that model, it implies no recombination taking place, but it differs in virtually all other respects from that model. If ultrafine structure represents induced reverse mutation rather than an extension of the genetic continuum, then the "map" it measures is the probability of such events occurring, not the distance between the pairs of mutant sites. Another possible explanation of ultrafine structure is that there are preferential sites of recombination, in which case the values obtained at very low frequency would not reflect the actual distance between sites.

A more serious aspect of fine structure exists at the "gross structure" level. Do the distances in pseudoallelic series represent intergenic or intra-

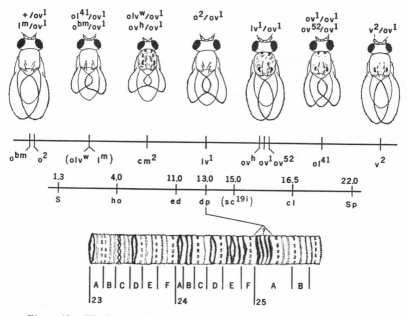

Figure 40. The Dumpy Region in Drosophila. It is not known whether this series is a single gene or an operon. Three different effects, on the wing (o), thorax (v), and viability (l) occur in different combinations. Their localization provides no meaningful pattern for complementation. Cytologically, the dumpy mutations resulting from chromosome breakage indicate that at least band 1 of the shoe buckle set in 25A affects the dumpy region. "Fine structure" apparently exists for the o and ov sub-regions. The mutants echinoid (ed) and clot (cl) are usually used as outside markers for recombination tests; (sc^{19i}) is the site at which the small y$^+$ac$^+$sc$^+$ region of the X chromosome was inserted into the second chromosome.

genic separations? For complex loci such as Notch, bithorax, and dumpy this is not known. In the lozenge series, where a serious attempt was made to obtain fine structure among seventeen different alleles, only three sites were obtained. Perhaps, if Green had extended his analysis to additional alleles with increased counts (about 100 fold greater than was used for obtaining his pseudoallelic distances) he might have obtained a fine structure in his allelic regions. It is also possible that this might be an "ultrafine structure" which is not related to crossing over. Definitive answers to this can be attempted in *Drosophila* where outside markers, rather than frequency, determine the existence of recombination and the placement of the alleles. In the dumpy series, the localized alleles are represented below:

$$o^{bm} \quad o^2 \quad \overset{\displaystyle olv^w}{l^m} \quad\quad cm^2 \quad lv^1 \quad ov^h \quad ov^1 \quad ov^{52b} \quad ol^{41} \quad v^2$$

The alleles o^{bm} and o^2 differ quantitatively from one another; so do the alleles ov^h and ov^1. Neither of these cases of recombination, found by Southin, is a critical demonstration of fine structure because the alleles used are perceptibly different. However, R. Sederoff has found recombination between ov^{52b} and ov^1, with ov^{52b} to the right of ov^1. These two alleles are indistinguishable in phenotype. This suggests that fine structure does exist in Drosophila; at least it rules out the pseudoallelic nature of the ov region of the dumpy system, whose members, six in all, are located between lv^1 and v^2. The frequency of these events, in both Sederoff's and Southin's cases, is about 1×10^{-5} which is considerably below the frequency of crossing over normally encountered in pseudoallelic series.

Should the ultrafine structure turn out to be a different phenomenon from fine structure, studies of gene conversion in fungi and higher organisms may require reconsideration. The marker placement test should be introduced into phage genetics; higher levels of resolution and more members of an allelic series are needed for Drosophila work. In each system the critical tests for interpreting these phenomena have not been done.

Position Effect and the Operon

The variegated position effects associated with heterochromatic rearrangements differ in two ways from the typical regulatory mechanisms inferred for the operon. The suppression of the mutant effect by additional heterochromatin makes it unlikely that this is an instance of the physical separation of structural genes from an operator. More than transcription appears to be involved. There is also a pattern of mutant and non-mutant segments of tissue which are apparently switched to "on" and "off" conditions in the normal development of these rearrangements. Stable position effects, by contrast, are not variegated and they are not repressed by heterochromatin. It is also difficult to find suitable analogies between the Bar case and the operon model. If the duplication separates a transcribing segment from its operator, why should double Bar be different in phenotype from Bar?

Additionally, little is known about the mechanisms of the spreading effect which affects apparently unrelated genes over a considerable portion of the chromosome. In some instances, a skipping effect is also noticeable—neighboring genes which manifest the position effect may surround a gene whose function is normal. These features of position effect reflect a looser coordination than that encountered in the operon.

Even in microbial systems the operon is not universally found for biochemical pathways. Some species of bacteria lack an operon organization for the pathways studied in Salmonella typhimurium and E. coli. In the fungi there is considerable distribution of the structural genes affecting a biochemical pathway; the genes are frequently found in different chromosomes. In such cases coordinated activity (such as repression and induction) would require an interchromosomal mechanism or non-genetic feedback controls.

Whether an operon oragnization does exist in *Drosophila* pseudoallelic systems or not, the concept is valuable because it focuses genetic control mechanisms on developmental problems. Unfortunately the biochemical approaches to the operon model are not available in most of the genetically favorable series in *Drosophila*. Eventually the application of this model to the problems of morphogenetic movements and tissue differentiation will be made. In man, the formation of the hemoglobin molecule in infants and adults may represent such an instance of regulatory control. However, until the necessary genetic tests can be made, the case for the operon in higher forms rests on analogy.

There are features of position effect which can be explored experimentally. Is position effect extended undirectionally or, as in Offermann's model, from either direction? If it is undirectional, is this a locus-specific effect or is there a consistent direction for the chromosome (e.g., distal to the centromere)? Are the physiological effects of temperature and crowding general or locus-specific effects? If breakage occurs in pseudoallelic regions, which components remain functional after the separation introduced by rearrangement? In this last instance, E. B. Lewis obtained a rearrangement in the bithorax region which permitted a partial functioning of some of the pseudoalleles. This case stands alone, but its value for analysis of the operon concept in *Drosophila* should be emphasized to prevent its lapsing into another instance of "research eclipse."

Chromosome Structure

The smallest known virus, $\phi_\chi 174$, contains one continuous DNA molecule with no free ends. This DNA is composed of 5500 nucleotides which are not paired in the mature, infectious particles but which do form a typical duplex with complementary bases in the vegetative, replicative form. The circularity of this virus DNA has been observed in electron micrographs. A circular organization of DNA also exists in certain bacteria. Kleinschmidt has developed a technique for the surface tension dispersion of DNA from ruptured bacterial cells; the electron micrographs reveal no free ends and in some cases total circularity can be seen. In T4 bacteriophage the genetic map is circular but the physical organization of the genetic material is still open to investigation. The circularity may represent linear molecules whose tips are homologous. G. Streisinger and F. Stahl refer to this condition as "terminal redundancy." A synaptic mechanism would be required for physical circularity if the redundant tips are duplex strands; if the tips are single-stranded and complementary then a true circularity would be achieved, but an ad hoc mechanism would be needed to account for the single-stranded state of the tips at the time of their formation.

The chromosomes of higher organisms are difficult to remove from cells without the physical or chemical destruction of their organization. Their massive size prevents single sections of embedded cells from giving unambiguous electron micrographs. The multiple levels of coiling in chromosomes also makes the continuity and strandedness of the chromosomes confused after sectioning.*

In studying strandedness, the ideal chromosomal state should be haploid. In bacteria, however, polynucleate conditions are common and it it difficult to obtain conditions which provide a colony containing uninucleate cells; it is also difficult to demonstrate that only one genome is present per cell. In higher organisms, somatic tissues may be polytene, or replication of the chromatids may have taken place prior to cellular differentiation. Thus cytological demonstrations of multiple chromatids in the chromosome of somatic tissues may have no bearing on the organization of the germinal chromosomes. The genetic analysis of gametic chromosomes often suggests the presence of a chromonema consisting of a single Watson-Crick duplex. This would not contradict the observation that the chromonemata of somatic chromosomes contain two or more DNA duplex threads.

The linear organization of microbial chromosomes, whether they have free ends or not, is that of a continuum. Exceptions to this may involve shearing effects and other traumatic artifacts which could give an illusion of complexity where none exists. Nevertheless the physical discontinuity of the chromosome in Drosophila, demonstrated by the left-right test, remains unchallenged. Its extension to other organisms and other regions of Drosophila would be valuable for a critical study of this problem. There is also another approach which has been eclipsed because of lack of interest. During the early 1950's a number of studies on chromosome fragility were made with chelating agents. These agents were reported to fragment the chromosomes into numerous small lengths of DNA, about the size of a few genes, as estimated by salivary chromosome analysis. Presumably the chelating agents affected calcium ion binding of these segments. These studies, however, were done before the shearing forces of extraction on DNA purification were known; such a possibility makes the interpretation of the fragments ambiguous. A repetition of chelating agent treatment on non-sheared DNA would be worthwhile for studies of the linear continuity and discontinuity of the chromosome.

---

* Some success in isolating DNA from the nuclei of wheat and mammals has been reported by Y. Hotta and A. Bassel, 1965, Proc. Nat. Acad. Sci. 53:356–362. Fragments of DNA up to 50 μ in length were obtained by extraction analysis and by visual observation of electron micrographs.

Point Mutation

The Watson-Crick model of DNA structure and Benzer's fine-structure analysis of the rII region in phage have provided the molecular basis for theories of mutagenesis which are likely to resolve Stadler's paradox. For some mutagens it is no longer necessary to consider point mutation as a residual class of undefined mutational events. The use of base analogues has made viral and bacterial mutation free of detectable rearrangements and consistent with predictions generated by their pairing specificities with the purine and pyrimidine bases in the replicating duplex. Direct tests of incorporation of analogues have been made biochemically, and some of the predicted pairing associations have been found.

The initial theories of chemical mutagenesis in bacteriophage are not unequivocally proved. The mechanism proposed for base analogues by E. Freese is the most widely accepted. This theory of "transition" mutations assumes, for example, that a purine analogue has a high probability of pairing with one of the two pyrimidines and a very low probability of pairing with the other pyrimidine. If the low-probability pairing occurs at the time of the incorporation of the analogue, the mutation will become fixed within two replications of the DNA. If the incorporation of the analogue occurs with the high-probability base, then many replications may elapse between the time of incorporation and the eventual "mispairing" of the analogue in the molecule with the low-probability base. The mutations which are obtained from the altered strands are revertible by base analogues.

On less secure grounds is a theory of "transversions" proposed by Freese for mutations induced by acridines. These acridine-induced mutations, which are not revertible by base analogues, were thought to be substitutions of purines for pyrimidines (or the reverse) at the mutant sites. This was rejected by the Cambridge school of Crick and Brenner, who argued that the acridines produced minute intragenic deletions or duplications, probably not larger than one nucleotide pair.

More difficulties exist in the interpretations of the mutational activity of such agents as nitrous acid, formaldehyde, and alkylating agents, all of whose reactions with DNA have turned out to be complex. Despite these limitations, these theories and approaches have given new stimulus to the research impasse reached on point mutation in higher organisms. It should now be possible to attempt a comparative mutagenesis in which the same agents may be used for microbial and multicellular systems. How striking the similarities and differences will be in the mutation process is still a matter of speculation.

Phylogenetic Study of the Genome

The techniques for isolating, purifying, and obtaining the molecular

TABLE 1.   *Estimated Number of Genes in Different Organisms**

ORGANISM	NUCLEOTIDE PAIRS	AVERAGE NO. OF GENES	T4 GENOME EQUIVALENTS
$\phi\chi 174$	5500	12	0.03
T4	2×10^5	450	1
E. coli	6×10^6	13,300	30
D. melanogaster	6×10^6	13,300	30
Cattle (sperm)	3×10^9	6.7×10^6	1.5×10^4

* The average gene size is based on a protein of 150 amino acids; if the codon is a triplet, the nucleotide pairs required would be 450.

weight of DNA from viruses and bacteria are excellent. In higher organisms, isolated nuclei may be analyzed by the same techniques used in phage and bacteria. In addition to these chemical methods, there are other techniques which can be used to obtain estimates of the DNA content of organisms. In viruses and bacteria the electron micrographs of ruptured cells or viral particles yield strands of DNA whose length corresponds to the calculations based on biochemical techniques. Similarly, in salivary gland chromosomes the length of the stretched X chromosomes (414 μ) is one fifth the length of the total genome. If this length represents the length of one of the "uncoiled" chromonemata in the salivary chromosome and if it is in the form of a Watson-Crick duplex, then the number of nucleotide pairs may be inferred from the model. This only requires the information that there are 10 nucleotide pairs per turn in the duplex and each turn has a length of 34 A units. In Table 1 the estimates for several organisms have been made. For the smallest organism, $\phi\chi 174$, the 5500 nucleotides present in the replicating form can encode 12 genes whose protein products are the size of an α or β chain of hemoglobin (about 150 amino acids, coded by 450 nucleotide pairs). T4 has about 35 times more DNA which gives it some 450 genes of this magnitude. In E. coli the genome is about 30 times larger than T4, providing it with some 13,000 genes. Surprisingly, the haploid genome of Drosophila would have the same number of genes as E. coli! But in the mammals the number of genes would be about 1000 times as great as in the fruit fly, with cattle or human sperm containing several million genes of this size.

There are several ways to interpret this. For the skeptical the techniques may be questioned. If the techniques are valid, however, then the estimates for Drosophila might be considered too low and the estimates for mammals too high. In the fruit fly calculations, the salivary chromosome might be considered insufficiently stretched. But if this were true then the number of genes in the scute-19 region would have to be increased in proportion to the amount of stretching assumed necessary. This would not likely be more than a two- or threefold stretch, however, when the length of the anaphase somatic chromosomes is compared to the enormous length of the corresponding salivary chromosomes. It is not likely that the average size of the gene is much smaller than that proposed for the α or β chain of hemoglobin. Table 2 shows the genes in the sc-19 region to have a size compatible with this molecule; if there were many more genes in this region, their average size would yield proteins with only a few tens of amino acids. If present estimates of the gene size in sc-19 are too low, and the region should be stretched, then, in the absence of invoking additional genes for this region, the number of amino acids per protein would be

TABLE 2. *Calculations of Gene Size and Numbers*

A. Assumptions Based on Watson-Crick DNA Duplex

1. 10 n. p. = 1 turn
2. 294 turns = 1 μ (34 A = 1 turn; 10,000 A = 1 μ)
3. 2940 n. p. = 1 μ
4. 450 n. p. = 150 amino acids for "average gene" (codon = 3)
5. 1 μ = 2940 n. p. = 6.5 average genes
6. Diameter of DNA = 12 Å

B. Size of Genome in *Drosophila melanogaster*

1. X chromosome = 414 μ
2. Total length = 5 X
3. $\dfrac{10 \text{ n. p.}}{1 \text{ turn}} \cdot \dfrac{294 \text{ turns}}{1 \mu} \cdot \dfrac{414 \mu}{1 \text{ X}} \cdot \dfrac{5 \text{ X}}{\text{genome}} = 6 \times 10^6$ n. p.
4. 6×10^6 n. p. = 13,300 average genes

C. Size of Gene: 1920 (1/30 μ, cubed) in Anaphase Chromosome

1. Length of cube = 1/30 μ
2. Nucleotide pairs per length $= \dfrac{1}{30} \mu \cdot \dfrac{10 \text{ n. p.}}{1 \text{ turn}} \cdot \dfrac{294 \text{ turns}}{1 \mu} = 98$ n. p.
3. Diameters per length $= \dfrac{1}{30} \mu \cdot \dfrac{10,000 \text{ Å}}{1 \mu} \cdot \dfrac{1 \text{ DNA molecule}}{12 \text{ Å}}$
   $= 27$ DNA molecules
4. 1/30 μ, cubed $= \dfrac{98 \text{ n. p.}}{\text{length}} \cdot \dfrac{27 \text{ molecules}}{\text{width}} \cdot \dfrac{27 \text{ molecules}}{\text{height}}$
   $= 72,000$ n. p.
5. 72,000 n. p. = 150 average genes

D. Size of Gene in Scute-19 Segment

1. Scute-19 = 0.5 μ
2. Scute-19 = 4 genes
3. $\dfrac{0.5 \mu}{1 \text{ sc-19}} \cdot \dfrac{1 \text{ sc-19}}{4 \text{ genes}} \cdot \dfrac{294 \text{ turns}}{1 \mu} \cdot \dfrac{10 \text{ n. p.}}{\text{turn}} = 367$ n. p.
4. 367 n. p. = 125 amino acids per gene

TABLE 2 (Continued)

E. Target Theory Genes

1. Volume of DNA $= 0.7854$ d^2l, where $d = 12$ Å
    $= 113$ l

2. Volume of target sphere $= 0.5236$ d^3
    $d_1 = 600$ Å (Blackstone), $v_1 = 113 \times 10^6$ Å^3
    $d_2 = 400$ Å (Gowen), $v_2 = 33.5 \times 10^6$ Å^3
    $d_3 = 260$ Å (Delbrück), $v_3 = 9.2 \times 10^6$ Å^3
    $d_4 = 20$ Å (Lea), $v_4 = 4.2 \times 10^3$ Å^3

3. $v_1/113 = 100$ $\mu = l_1$
    $v_2/113 = 30$ $\mu = l_2$
    $v_3/113 = 9$ $\mu = l_3$
    $v_4/113 = 38$ Å $= l_4$

4. 100 $\mu = 6.5 \times 10^3$ average genes $=$ Blackstone's gene
    30 $\mu = 195$ average genes $=$ Gowen's gene
    9 $\mu = 58$ average genes $=$ Delbrück's gene
    38 Å $= 10$ nucleotide pairs $=$ Lea's gene

several hundred. A tenfold increase, for example, would lead to massive proteins for these genes.

If the *Drosophila* estimates are too low, what can be said for the mammalian genome? Is only a small fraction of the mammalian DNA involved in the coding of structural genes? Are the chromosomes of mammalian gametes almost as polytene as salivary chromosomes? Is only a small fraction of the mammalian DNA involved in the coding of structural genes and virtually all the rest of it manifested through regulatory functions? This is hard to believe if the complexity of multicellular organization is achieved in *Drosophila* with the same number of genes as exists in bacteria!*

In the future, genetic mapping may include regions of regulatory control with no known enzymatic or structural protein product. Until then these considerations on the comparative sizes of genes among unrelated forms will remain enigmatic. In addition to the genomes themselves, I have attempted to convert the size of the gene in *Drosophila* into nucleotide pairs. Table 2 shows how much difference exists in the various models that have been proposed over the past fifty years. Note that the gene based on the segment of an anaphase chromosome is about ten times larger than the entire genome of $\phi\chi174$. However, somatic (possibly polytene)

---

* Muller has suggested (see reference, p. vii) that the genes in higher animals form giant proteins whose amino acid sequence contains ten to one hundred times as many residues as the average microbial protein. Another alternative is that purified spermatozoa contain substantial amounts of *mitochondrial* DNA in the midpiece. If so, more than 90 percent of sperm DNA would be mitochondrial, assuming the average size gene to be identical to that found in microbes.

chromosomes were used in this estimate and it is thus not a very reliable approach. The sc-19 segment with its four genes gives an average size of about 100 amino acids for each protein. The target theory genes (with one exception) are *not* smaller than genes coding for a protein with 150 amino acid residues. The estimates of Blackwood, Delbrück, and Gowen are several times larger than that! As Muller pointed out, however, any correspondence of the size of the sensitive target with the gene itself would only be fortuitous because so many fallacies and unproved assumptions are used for this theory. If, as seems likely, the genetic code is universal, these phylogenetic differences in DNA content must be resolved before the role of the code in the genome can be understood.

At the turn of the century Bateson proclaimed that not the slightest thing was known about the mechanisms of heredity. In sixty-five years the concept of the gene has matured from a hypothetical unit. The location of the gene in the chromosome and its mode of transmission were the immediate concern of the pioneers of genetics. Its modification through mutation represented the second phase of study, leading to new techniques and the use of new organisms for explaining the problems of gene size and function. The molecular phase, with its suggestive models of coding and regulation are preparing the way for a new phase of genetics which will occupy the talents of scientists in the closing decades of the twentieth century—the role of genetics in the organization of the living cell and in the coordinated activities of cells which represent the living organism.

REFERENCES

1. AGOL, I. J. 1929. "Stepallelomorphism in Drosophila melanogaster." Zhurnal eksperimental 'noi i moditsins 5:86–101. (PST Cat. No. 518 transl.: Israel Program for Scientific Translations.)
2. ALTENBURG, E. 1946. Commentary on T. M. Sonneborn's paper. Cold Spring Harbor Symposia on Quantitative Biology 11:236–255.
3. ALTENBURG, E. 1948. "The Role of Symbionts and Autocatalysts in the Genetics of the Ciliates." American Naturalist 82:252–264.
4. AMES, B. N., and B. GARRY. 1959. "Coordinate Repression of the Synthesis of Four Histidine Biosynthetic Enzymes by Histidine." Proceedings of the National Academy of Science 45:1453–1461.
5. AVERY, O. T., C. M. MACLEOD, and M. MCCARTY. 1944. "Studies on the Chemical Nature of the Substance Inducing Transformation of Pneumococcal Types." Journal of Experimental Biology and Medicine 79:137–158.
6. BATESON, W. 1897. "On Progress in the Study of Variation." Science Progress, n. s. Vol. 1, no. 5:1–15.
7. BATESON, W. 1898. "On Progress in the Study of Variation," Part II. Science Progress, n. s. Vol. 2, no. 6:1–16.
8. BATESON, W. 1900. "Hybridization and Cross-breeding as a Method of Scientific Investigation." Journal of the Royal Horticultural Society 24:1–8.
9. BATESON, W. 1900. "Problems of Heredity as a Subject for Horticultural Investigation." Journal of the Royal Horticultural Society 25:1–8.
10. BATESON, W. 1900. "Problems of Heredity and Their Solution." Journal of the Royal Horticultural Society 25:54–61.
11. BATESON, W. 1901. "Heredity, Differentiation, and Other Conceptions of Biology: a Consideration of Professor Karl Pearson's Paper 'on the Principle of Homotyposis.'" Proceedings of the Royal Society 69:193–205.
12. BATESON, W. 1902. A Defence of Mendel's Principles of Heredity. Cambridge University Press, Cambridge.

273

13. BATESON, W. 1903. "Variation and Differentiation in Parts and Breth-ren." Privately printed, Cambridge.

14. BATESON, W. April 18, 1905, letter to A. Sedgwick, from: Bateson, William, 1928, *Essays and Addresses*, edited by B. Bateson. Cambridge University Press, Cambridge.

15. BATESON, W. 1906. Presidential report, "Progress of Genetic Research." *3rd International Conference of Genetics*, pp. 90–97.

16. BATESON, W. 1906. Inaugural address, "The Progress of Genetic Research," *3rd Conference on Hybridization and Plant Breeding*, pp. 90–97.

17. BATESON, W. 1906. "The Progress of Genetics Since the Rediscovery of Mendel's Papers." From *Progressus rei Botanicae, Assoc. Internat. des Botanites*, edited by J. P. Lotsy, pp. 368–418, G. Fischer, Jena.

18. BATESON, W. 1908. "The Methods and Scope of Genetics," p. 22. Cambridge University Press, Cambridge.

19. BATESON, W. 1910. "Sketch of Brook's Life by Some of His Former Students and Associates." *Journal of Experimental Zoology* 9:1–52.

20. BATESON, W. 1913 (address delivered in 1907). "Problems of Genetics." Yale University Press, New Haven, Conn.

21. BATESON, W. 1919. "Science and Nationality." *Edinburgh Review* 229: 123–138.

22. BATESON, W. 1922. "Evolutionary Faith and Modern Doubts." *Science* 55:1412.

23. BATESON, W. 1926. "Segregation." *Journal of Genetics* 16:201–235.

24. BATESON, W. 1928. *Essays and Addresses*, biographical memoir, edited by Beatrice Bateson, pp. 28–73. Cambridge University Press, Cambridge.

25. BATESON, W., and R. C. PUNNETT. 1911. "On Gametic Series Involving Reduplication of Certain Terms." *Journal of Genetics* 1:239–302.

26. BATESON, W., and E. R. SAUNDERS. 1902. "Experimental Studies in the Physiology of Heredity." *Reports to the Evolution Committee of the Royal Society* 1:1–160.

27. BATESON, W., E. SAUNDERS and R. PUNNETT. 1904. "Report II: Experimental Studies in the Physiology of Heredity." *Reports to the Evolution Committee of the Royal Society* 2:1–154.

28. BATESON, W., E. SAUNDERS, and R. PUNNETT. 1906. "Report III: Experimental Studies in the Physiology of Sex." *Reports to the Evolution Committee of the Royal Society* 3:1–53.

29. BATESON, W., E. R. SAUNDERS, and R. C. PUNNETT. 1908. "Report IV: Experimental Studies in the Physiology of Heredity." *Reports to the Evolution Committee of the Royal Society* 4:1–60.

30. BEADLE, G. W. 1945. "The Genetic Control of Biochemical Reactions." *Harvey Lectures*, series 40:179–194.

31. BEADLE, G. W., and B. EPHRUSSI. 1935. "Transplantation in Droso-phila." *Proceedings of the National Academy of Science* 21:642–646.

32. BEADLE, G. W., and B. EPHRUSSI. 1936. "The Differentiation of Eye Pigments in Drosophila as Studied by Transplantation." *Genetics* 21:225–247.

33. BEADLE, G. W., and E. L. TATUM. 1941. "Genetic Control of Development and Differentiation." *American Naturalist* 75:107–116.

34. BEADLE, G. W., and E. L. TATUM. 1941. "Genetic Control of Biochemical Reactions in Neurospora." *Proceedings of the National Academy of Science* 27:499–506.

35. BELLING, J. 1928. "The Ultimate Chromomeres in Lilium and Aloë

with Regard to the Numbers of Genes." *University of California Publication of Botany* 14:307–318.

36. BELLING, J. 1931. "Chromomeres of Liliaceous Plants." *University of California Publication of Botany* 16:153–170.

37. BELOZERSKY, A. N., and A. S. SPIRIN. 1958. "A Correlation between the Compositions of Deoxyribonucleic and Ribonucleic Acids." *Nature* 182:111.

38. BENZER, S. 1955. "Fine Structure of a Genetic Region in Bacteriophage." *Proceedings of the National Academy of Science* 41:344–354.

39. BENZER, S. 1956. "Genetic Fine Structure and Its Relation to the DNA Molecule." *Brookhaven Symposia in Biology* 8:3–16.

40. BENZER, S. 1957. *The Chemical Basis of Heredity*, edited by W. D. McElroy and B. Glass, pp. 70–93. The Johns Hopkins Press, Baltimore, Md.

41. BENZER, S. 1959. "On the Topology of the Genetic Fine Structure." *Proceedings of the National Academy of Science* 45:1607–1620.

42. BENZER, S. 1961. "On the Topography of the Genetic Fine Structure." *Proceedings of the National Academy of Science* 47:403–416.

43. BLACKWOOD, O. 1931. "X-ray Evidence as to the Size of a Gene" (abstract). *Physical Revue* 37:1698.

44. BLACKWOOD, O. 1932. "Further X-ray Evidence as to the Size of a Gene; and as to the Energy of Mutation by Ultraviolet Rays" (abstract). *Physical Revue* 40:1034.

45. BRENNER, S. 1957. "On the Impossibility of All Overlapping Triplet Codes in Information Transfer from Nucleic Acid to Proteins." *Proceedings of the National Academy of Science* 43:687.

46. BRENNER, S., and L. BARNETT. 1959. "Genetic and Chemical Studies on the Head Protein of Bacteriophages T2 and T4." *Brookhaven Symposia in Biology* 12:86–94.

47. BRENNER, S., F. JACOB, and M. MESELSON. 1961. "An Unstable Intermediate Carrying Information from Genes to Ribosomes for Protein Synthesis." *Nature* 190:576–581.

48. BRENNER, S., L. BARNETT, F. H. C. CRICK, and A. ORGEL. 1961. "The Theory of Mutagenesis." *Journal of Molecular Biology* 3:121–124.

49. BRIDGES, C. B. 1913. "Non-disjunction of the Sex Chromosomes of Drosophila." *Journal of Experimental Zoology* 15:587–606.

50. BRIDGES, C. B. 1914. "Direct Proof through Non-disjunction that the Sex-linked Genes of Drosophila Are Borne by the X-chromosome." *Science*, n. s. 40:107–109.

51. BRIDGES, C. B. 1917. "An Intrinsic Difficulty for the Variable Force Hypothesis of Crossing Over." *American Naturalist* 51:370–373.

52. BRIDGES, C. B. 1917. "Deficiency." *Genetics* 2:445–465.

53. BRIDGES, C. B. 1919. "Specific Modifiers of Eosin Eye Color in Drosophila Melanogaster." *Journal of Experimental Zoology* 28:337–384.

54. BRIDGES, C. B. 1936. "The Bar 'Gene' a Duplication." *Science* 83:210–211.

55. BRINK, R. A. 1956. "A Genetic Change Associated with the R Locus in Maize which Is Directed and Potentially Reversible." *Genetics* 41:872–889.

56. BROOKS, W. K. 1899. *The Foundations of Zoology*. The Macmillan Co. (Columbia University Press), New York.

57. BROOKS, W. K. 1906. "Heredity and Variation—Logical and Biological." *Proceedings of the American Philosophy Society* 45:70–76.

58. CARLSON, E. A. 1959. "Allelism, pseudoallelism, and complementation at the dumpy locus in D. melanogaster." *Genetics* 44:347–373.
59. CARLSON, E. A. 1959. "Comparative Genetics of Complex Loci." *Quarterly Review of Biology* 34:33–67.
60. CARLSON, E. A. 1961. "Limitations of Geometrical Models for Complementation Mapping of Allelic Series." *Nature* 191:788–790.
61. CASE, M., and N. GILES. 1958. "Recombination Mechanisms at the Pan-2 Locus in Neurospora Crassa." *Cold Spring Harbor Symposia on Quantitative Biology* 23:119–135.
62. CASE, M. E., and N. H. GILES. 1960. "Comparative Complementation and Genetic Maps of the Pan-2 Locus in Neurospora Crassa." *Proceedings of the National Academy of Science* 46:659.
63. CASTLE, W. 1903. "Mendel's Law of Heredity." *Proceedings of the American Academy of Arts and Science* 38:535–548.
64. CASTLE, W. 1905. "Recent Discoveries in Heredity and Their Bearing on Animal Breeding." *Popular Science Monthly* 66:193–208.
65. CASTLE, W. 1906. "Yellow Mice and Gametic Purity," *Science, n. s.* 24:275–281.
66. CASTLE, W. E. 1910. "Heredity." *Popular Science Monthly* 71:417–428.
67. CASTLE, W. E. 1911. Heredity. D. Appleton and Co., New York.
68. CASTLE, W. E. 1912. "The Inconstancy of Unit-Characters." *American Naturalist* 46:352–362.
69. CASTLE, W. E. 1913. "Simplification of Mendelian Formulae." *American Naturalist* 47:170–182.
70. CASTLE, W. 1914. "Mr. Muller on the Constancy of Mendelian Factors." *American Naturalist* 49:37–42.
71. CASTLE, W. 1914. "Multiple Factors in Heredity." *Science, n. s.* 39:686–689.
72. CASTLE, W. E. 1914. "Pure Lines and Selection." *Journal of Heredity* 5:93–97.
73. CASTLE, W. E. 1916. "Can Selection Cause Genetic Change?" *American Naturalist* 50:248–256.
74. CASTLE, W. E. 1916. "Is Selection or Mutation the More Important Agency in Evolution." *Scientific Monthly* Jan. 1916; 91–98.
75. CASTLE, W. E. 1919. "Piebald Rats and the Theory of Genes." *Proceedings of the National Academy of Science* 5:126–130.
76. CASTLE, W. 1919. "Are Genes Linear or Non-linear in Arrangement?" *Proceedings of the National Academy of Science* 5:500–506.
77. CASTLE, W. 1919. "Is the Arrangement of the Genes in the Chromosome Linear?" *Proceedings of the National Academy of Science* 5:25–32.
78. CASTLE, W. 1920. "Model of the Linkage System of Eleven Second Chromosome Genes of Drosophila." *Proceedings of the National Academy of Science* 6:73–77.
79. CASTLE, W., and G. M. ALLEN. 1903. "Mendel's Law and the Heredity of Albinism." *Mark's Anniversary Volume*, article 19, pp. 379–398. Harvard University Press, Cambridge, Mass.
80. CASTLE, W. E., and G. M. ALLEN. 1903. "Heredity of Albinism." *Proceedings of the American Academy of Arts and Science* 38:603–622.
81. CASTLE, W., and P. B. HADLEY. 1915. "The English Rabbit and the Question of Mendelian Unit-character Constancy." *Proceedings of the National Academy of Science* 1:39–42.
82. CASTLE, W. E. and J. C. PHILLIPS. 1909. "A Successful Ovarian Trans-

plantation in the Guinea-pig, and Its Bearing on Problems of Genetics." *Science*, n. s. *30*:312–314.

83. CASTLE, W. E. and J. C. PHILLIPS. 1914. *Piebald Rats and Selection.* Carnegie Institute Washington Publication #195.

84. CATCHESIDE, D. G. 1964. "Interallelic Complementation." *Brookhaven Symposia in Biology 17*:1–14.

85. CATCHESIDE, D. G. and A. OVERTON. 1958. "Complementation between Alleles in Heterocaryons." *Cold Spring Harbor Symposia on Quantitative Biology 23*:137.

86. CONKLIN, E. G. 1898. *The Factors of Organic Evolution from the Standpoint of Embryology,* edited by D. S. Jordan. D. Appleton and Co., New York.

87. CONKLIN, E. G. 1905. "The Mutation Theory from the Standpoint of Cytology." *Science,* n.s. *21*:525–529.

88. CORRENS, C. 1919. "Vererbungsversuche mit buntblättrigen Sippen I." Sitzungsvor. Preuss. Ak. Wiss. *34.*

89. CRICK, F. H. C. 1958. "On Protein Synthesis." *Symposium of the Society of Experimental Biology 12*:138–167.

90. CRICK, F. H. C. 1959. "The Present Position of the Coding Problem." *Brookhaven Symposia in Biology 12*:35–39.

91. CRICK, F. H. C., and L. ORGEL. 1964. "The Theory of Inter-allelic Complementation." *Journal of Molecular Biology 8*:161.

92. CROWTHER, J. A. 1924. "Some Considerations Relative to the Action of X-rays on Tissue Cells." *Proceedings of the Royal Society of Biology 96*:207–211.

93. CUÉNOT, L. 1903. "L'hérédité de la pigmentation ches les souris." *Arch. de Zool. exper. et gen. 1* (4th s.):33–41.

94. DARLINGTON, C. D. 1944. "Heredity, Development, and Infection." *Nature 154*:164–169.

95. DARWIN, C. 1868. "Provisional Hypothesis of Pangenesis." *Animals and Plants Under Domestication,* Vol. II, pp. 428–483. Orange Judd & Co., New York.

96. DAVENPORT, C. B. 1908. "Determination of Dominance in Mendelian Inheritance." *Proceedings of the American Philosophy Society 47*:59–63.

97. DEMEREC, M. 1926. "Mutable Genes in Drosophila Virilis." *Proceedings of the International Congress of Plant Science (Ithaca) 1*:943–946.

98. DEMEREC, M. 1928. "Mutable Characters of Drosophila Virilis 1. Reddish-α Body Character." *Genetics 13*:359–388.

99. DEMEREC, M. 1928. "The Behavior of Mutable Genes." *Proceedings of the 5th International Congress of Genetics, 1927,* Suppl. 1, ZiAV 1928:183–193.

100. DEMEREC, M. 1931. "The Gene." *Biology Laboratory,* (Cold Spring Harbor, New York) *3*:29–32.

101. DEMEREC, M. 1933. "What Is a Gene?" *Journal of Heredity 64*:369–378.

102. DEMEREC, M. 1934. "The Gene and Its Role in Ontogeny." *Cold Spring Harbor Symposia on Quantitative Biology 2*:110–115.

103. DEMEREC, M. 1935. "Unstable Genes." *Botanical Review 1*:233–248.

104. DEMEREC, M. 1938. "Eighteen Years of Research on the Gene." *Cooperation in Research, Carnegie Institute of Washington Publication 501*:295–314.

105. DEMEREC, M. 1939. "Chromosome Structure as Viewed by a Geneticist." *American Naturalist 73*:331–338.

106. DEMEREC, M. 1955. "What Is a Gene—Twenty Years Later." *American Naturalist* 89:5–20.

107. DEMEREC, M., I. BLOMSTRAND, and Z. E. DEMEREC, 1955. "Evidence of Complex Loci in Salmonella." *Proceedings of the National Academy of Science* 41:359–364.

108. DEMEREC, M., and Z. DEMEREC. 1956. "Analysis of Linkage Relationships in Salmonella by Transduction Techniques." *Brookhaven Symposia in Biology* 8:75–87.

109. DEMEREC, M., et al. 1954. "Bacterial Genetics." *Annual Report, Department of Genetics, Carnegie Institute Washington Year Book* 53:225–241.

110. DE VRIES, H. 1889. *Intracellular Pangenesis*, translated by C. S. Gager, 1910. Open Court Publishers, Chicago.

111. DE VRIES, H. 1900. "Sur la loi de disjunction des Hybrides." *C. R. Acad. Sci. (Paris)* 130:845–847.

112. DE VRIES, H. 1903. "Fertilization and Hybridization" (read, in Dutch, at the 151st annual meeting of the Dutch Society of Science, Haarlem, May 16, 1903). Translated by C. S. Gager, 1910, in *Intracellular Pangenesis*, Open Court Publishers, Chicago.

113. DE VRIES, H. 1901–03. *Die Mutationstheorie.* Veit & Co., Leipzig.

114. DE VRIES, H. 1914. "Priniciples of the Theory of Mutation." *Science,* n. s. 40:77–84.

115. DOBZHANSKY, T. 1932. "The Baroid Mutation in Drosophila Melanogaster." *Genetics* 17:369–392.

116. DREYER, W. J. 1960. "Comment on C. A. Knight's paper." *Brookhaven Symposia in Biology* 13:243–248.

117. DUBININ, N. P. 1932. "Stepallelomorphism in *Drosophila melanogaster*— the allelomorphs achaete[2]-scute[10], achaete[1]-scute[11], and achaete[3]-scute[13]." *Journal of Genetics* 25:161–181.

118. EAST, E. M. 1910. "A Mendelian Interpretation of Variation that is Apparently Continuous." *American Naturalist* 44:65–82.

119. EAST, E. M. 1912. "The Mendelian Notation as a Description of Physiological Facts." *American Naturalist* 46:633–695.

120. EAST, E. M. 1923. "Mendel and His Contemporaries." *Science Monthly* 16:225–236.

121. EAST, E. M. 1929. "The Concept of the Gene." *Proceedings of the International Congress of Plant Science* 1:889–895.

122. EDGAR, R. S., G. H. DENHART, and R. H. EPSTEIN. 1964. "A Comparative Genetic Study of Conditional Lethal Mutations of Bacteriophage T4D." *Genetics* 49:635–648.

123. EMERSON, R. A. 1911. "Genetic Correlation and Spurious Allelomorphism in Maize." *24th Annual Report, Nebraska Agricultural Experimental Station*, pp. 59–90.

124. EMERSON, R. A. 1917. "Genetical Studies of Variegated Pericarp in Maize." *Genetics* 2:1–35.

125. EMERSON, R. A. 1929. "The Frequency of Somatic Mutation in Variegated Pericarp of Maize." *Genetics* 14:488–511.

126. EMERSON, R. A., W. H. EYSTER, E. G. ANDERSON, and M. DEMEREC. 1922. "Studies of Somatic Mutations in Variegated Maize Pericarp." *Anatomical Record* 23:90–91.

127. EYSTER, W. H. 1924. "A Genetic Analysis of Variegation." *Genetics* 9:372–404.

128. FINCHAM, J. R. S., and J. A. PATEMAN. 1957. "Formation of an Enzyme

through Complementary Action of Mutant Alleles in Separate Nuclei in a Heterocaryon." *Nature* 179:741–742.

129. FREESE, E. 1959. "On the Molecular Explanation of Spontaneous and Induced Mutations." *Brookhaven Symposia in Biology* 12:63–75.

130. GALTON, F. 1886. "Hereditary Stature." *Nature* 33:295–298.

131. GALTON, F. 1897. "The Average Contribution of Each Several Ancestor to the Total Heritage of the Offspring." *Proceedings of the Royal Society* 61:401–413.

132. GAMOW, G. 1954. "Possible Relation between Deoxyribonucleic Acid and Protein Structure." *Nature* 173:318.

133. GARROD, A. E. 1908. "Inborn Errors of Metabolism." *Lancet*, July 4 issue, 1–7.

134. GATES, R. R. 1915. "On the Nature of Mutations." *Journal of Heredity* 6:99–108.

135. GATES, R. R. 1925. "Mutation." *Nature* 115:499–500.

136. GILES, N. 1958. "Comment in *Discussion* of Catcheside and Overton's paper." *Cold Spring Harbor Symposia on Quantitative Biology* 23:140.

137. GILES, N. H., C. W. H. PARTRIDGE, and N. J. NELSON. 1957. "The Genetic Control of Adenylosuccinase in *Neurospora crassa*." *Proceedings of the National Academy of Science* 43:305–517.

138. GLASS, B. 1933. "A New Allelomorphic Compound Presenting the Phenotype of the Wild *Drosophila melanogaster*." *Journal of Genetics* 27:233–241.

139. GLASS, B. 1953. "The Long Neglect of a Scientific Discovery: Mendel's Laws of Inheritance." *Studies in Intellectual History*. The Johns Hopkins Press, Baltimore, Md.

140. GOLDSCHMIDT, R. 1916. "Genetic Factors and Enzyme Reaction." *Science*, n. s. 43:98–100.

141. GOLDSCHMIDT, R. 1917. "A Preliminary Report on Some Genetic Experiments Concerning Evolution." *American Naturalist* 52:28–50.

142. GOLDSCHMIDT, R. 1917. "Crossing-over ohne Chiasmatypie?" *Genetics* 2:82–95.

143. GOLDSCHMIDT, R. 1928. "The Gene." *Quarterly Review of Biology* 3:307–324.

144. GOLDSCHMIDT, R. 1932. "Genetics and Development." *Biology Bulletin* 63:337–356.

145. GOLDSCHMIDT, R. 1934. "The Influence of the Cytoplasm upon Gene-controlled Heredity." *American Naturalist* 68:5–23.

146. GOLDSCHMIDT, R. 1937. "Spontaneous Chromatin Rearrangements in Drosophila." *Nature* 140:767.

147. GOLDSCHMIDT, R. 1938. "The Theory of the Gene." *Science Monthly* 46:268–273.

148. GOLDSCHMIDT, R. 1940. "Chromosomes and Genes." AAAS publication 14:56–66.

149. GOLDSCHMIDT, R. B. 1944. *Science in the University*. University of California Press, Berkeley.

150. GOLDSCHMIDT, R. B. 1946. "Position Effect and the Theory of the Corpuscular Gene." *Experientia* 2:1–40.

151. GOLDSCHMIDT, R. 1950. " 'Repeats' and the Modern Theory of the Gene." *Proceedings of the National Academy of Science* 36:365–368.

152. GOLDSCHMIDT, R. 1950. "Fifty Years of Genetics." *American Naturalist* 84:313–340.

153. GOLDSCHMIDT, R. 1954. "Different Philosophies of Genetics." Science 119:703–710.
154. GOWEN, J. W. 1933. "Meiosis as a Genetic Character in Drosophila melanogaster." Journal of Experimental Zoology 65:83–106.
155. GOWEN, J. W., and E. H. GAY. 1932. "Eversporting as a Function of the Y-chromosome in Drosophila Melanogaster." Proceedings of the National Academy of Science 19:122–126.
156. GOWEN, J. W., and E. H. GAY. 1933. "Effect of Temperature on Eversporting Eye Color in Drosophila Melanogaster." Science 77:312.
157. GOWEN, J. W., and E. H. GAY. 1933. "Gene Number, Kind and Size in Drosophila." Genetics 18:1–31.
158. GREEN, M. M. 1959. "Spatial and Functional Properties of Pseudoalleles at the White Locus in Drosophila Melanogaster." Heredity 13: 302–315.
159. GREEN, M. M., and K. C. GREEN. 1956. "A Cytogenetic Analysis of the Lozenge Pseudoalleles in Drosophila." Zeitschrift für induktiv Vererbungslehre 87:708–721.
160. GROSS, S. R. 1962. "On the Mechanism of Complementation at the leu-2 Locus of Neurospora." Proceedings of the National Academy of Science 48:922–930.
161. HALDANE, J. B. S. 1930. "A Note on Fisher's Theory of the Origin of Dominance, and on a Correlation between Dominance and Linkage." American Naturalist 64:87–90.
162. HARRISON, R. G. 1937. "Embryology and Its Relations." Science 85:369–374.
163. HAYASHI, M., M. N. HAYASHI, and S. SPIEGELMAN. 1964. "DNA Circularity and the Mechanism of Strand Selection in the Generation of Genetic Messages." Proceedings of the National Academy of Science 51:351–359.
164. HURST, C. 1906. "Mendelian Characters in Plants and Animals." Report of the 3rd International Conference of Genetics, 114–128.
165. HYDE, R. R. 1915. "On the Change that Takes Place in the Chromosome in Mutating Stocks." Proceedings of the Indiana Academy of Science 1915:339–344.
166. HYDE, R. R. 1916. "Two New Members of a Sex-linked Multiple (Sextuple) Allelomorph System." Genetics 1:535–580.
167. HYDE, R. R. 1920. "Segregation and Recombination of the Genes for Tinged, Blood, Buff, and Coral in Drosophila Melanogaster." Proceedings of the Indiana Academy of Science 1920:291–300.
168. INGRAM, V. M. 1956. "A Specific Chemical Difference between the Globins of Normal Human and Sickle-cell Anemia Haemoglobin." Nature 178:792–794.
169. JACOB, F., and J. MONOD. 1959. "Gènes de structure and gènes de régolution dans la biosynthèse des protéines." C. R. Acad. Sci. (Paris) 249:1282.
170. JACOB, F., and J. MONOD. 1961. "Genetic Regulatory Mechanisms in the Synthesis of Proteins." Journal of Molecular Biology 3:318–356.
171. JACOB, F., and J. MONOD. 1961. "On the Regulation of Gene Activity." Cold Spring Harbor Symposia on Quantitative Biology 26:193–209.
172. JACOB, F., D. PERRIN, C. SANCHEZ, and J. MONOD. 1960. "L'opéron: groupe de gènes à expression coordonnée par un opérateur." C. R. Acad. Sci. (Paris) 250:1727–1729.
173. JANSSENS, F. A. 1909. "La Theorie de la Chiasmatypie." La Cellule 25: 389.

174. JOHANNSEN, W. 1906 (July 31). "Does Hybridisation Increase Fluctuating Variability?" *Report of the 3rd International Congress of Genetics*, pp. 98–112.

175. JOHANNSEN, W. 1909. "Elemente der Exakten Erblichkeitslehre." G. Fischer, Jena.

176. KAPULER, A. M., and H. BERNSTEIN. 1963. "A Molecular Model for an Enzyme Based on a Correlation between the Genetic and Complementation Maps of the Locus Specifying the Enzymes." *Journal of Molecular Biology* 6:443–451.

177. LEA, D. E. 1947. *Actions of Radiations on Living Cells*. Cambridge University Press (The Macmillan Co.), New York.

178. LEDER, P., and M. W. NIRENBERG. 1964. "RNA Codeword and Protein Synthesis; III on the Nucleotide Sequence of a cyst. and leu. RNA Codeword." *Proceedings of the National Academy of Science* 51: 1521–1529.

179. LEDERBERG, J. 1957. *Chemical Basis of Heredity*. The Johns Hopkins Press, Baltimore, Md.

180. LEWIS, E. B. 1941. "Another Case of Unequal Crossing Over in Drosophila Melanogaster." *Proceedings of the National Academy of Science* 27:31–34.

181. LEWIS, E. B. 1942. "The Star and Asteroid Loci in Drosophila Melanogaster." *Genetics* 27:153–154.

182. LEWIS, E. B. 1945. "The Relation of Repeats to Position Effect in Drosophila Melanogaster." *Genetics* 30:137–166.

183. LEWIS, E. B. 1948. "Pseudoallelism in Drosophila Melanogaster." *Genetics* 33:113.

184. LEWIS, E. B. 1951. "Pseudoallelism and Gene Evolution." *Cold Spring Harbor Symposia on Quantitative Biology* 16:159–174.

185. LEWIS, E. B. 1952. "The Pseudoallelism of White and Apricot in Drosophila Melanogaster." *Proceedings of the National Academy of Science* 38:953–961.

186. LEWIS, E. B. 1954. "Caryologia." *Proceedings of the 9th International Congress of Genetics*, pp. 100–105.

187. LEWIS, E. B. 1955. "Some Aspects of Position Pseudoallelism." *American Naturalist* 89:73–89.

188. LOCK, R. H. 1906. *Recent Progress in the Study of Variation, Heredity, and Evolution*. J. Murray, London.

189. MACKENDRICK, M. E., and G. PONTECORVO. 1952. "Crossing-over between Alleles at the W Locus in Drosophila Melanogaster." *Experientia* 8:390.

190. MAY, H. G. 1917. "The Appearance of Reverse Mutations in the Bar-eyed Race of *Drosophila* under Experimental Control." *Proceedings of the National Academy of Science* 3:544–545.

191. McCLINTOCK, B. 1956. "Controlling Elements and the Gene." *Cold Spring Harbor Symposia on Quantitative Biology* 21:197–216.

192. MENDEL, G. 1965. "Experiment in Plant-hybridization." *Verh. naturf. Ver. in Brunn*, Abv. iv. 1865.

193. MONOD, J., J. P. CHANGEUX, and F. JACOB. 1963. "Allosteric Proteins and Cellular Control Systems." *Journal of Molecular Biology* 6: 306–329.

194. MONOD, J., and G. COHEN-BAZIRE. 1953. "L'effet d'inhibition spécifique dans la biosynthèse de la tryptophane-desmase chez Aerobacter." *C. R. Acad. Sci. (Paris)* 236:530.

195. MORGAN, T. H. 1901. "The Problem of Development." *International Monthly*, March 1901.
196. MORGAN, T. H. 1905. "The Origin of Species." *Popular Science Monthly* 67:54–65.
197. MORGAN, T. H. 1909. "For Darwin." *Popular Science Monthly*, April 70:367–380.
198. MORGAN, T. H. 1910. "Chromosomes and Heredity." *American Naturalist* 44:449–496.
199. MORGAN, T. H. 1910. "Sex Limited Inheritance in Drosophila." *Science* 32:120–122.
200. MORGAN, T. H. 1911. "Chromosomes and Associative Inheritance." *Science, n. s.* 34:636–638.
201. MORGAN, T. H. 1911. "Random Segregation Versus Coupling in Mendelian Inheritance." *Science* 34:384.
202. MORGAN, T. H. 1912. "Complete Linkage in the Second Chromosome of the Male of Drosophila." *Science* 36:719–720.
203. MORGAN, T. H. 1912. "The Explanation of a New Sex Ratio in Drosophila." *Science* 36:718–719.
204. MORGAN, T. H. 1914. "The Mechanism of Heredity as Indicated by the Inheritance of Linked Characters." *Popular Science Monthly*, Jan. 1914, 1–16.
205. MORGAN, T. H. 1914. "The Failure of Ether to Produce Mutations in Drosophila." *American Naturalist* 48:705–711.
206. MORGAN, T. H. 1915. "The Constitution of the Hereditary Material." *Proceedings of the American Philosophy Society* 54:143–153.
207. MORGAN, T. H. 1917. "The Theory of the Gene." *American Naturalist* 51:513–544.
208. MORGAN, T. H. 1918. "Changes in Factors through Selection." *Science Monthly*, June 1918, 549–559.
209. MORGAN, T. H. 1922. "On the Mechanism of Heredity." *Proceedings of the Royal Society of Biology* 94:162–197.
210. MORGAN, T. H. 1932. "The Rise of Genetics." *Science* 76:261–267, 285–288.
211. MORGAN, T. H. 1939. "Personal Recollections of Calvin B. Bridges." *Journal of Heredity* 30:355.
212. MORGAN, T. H. 1940. "Obituary Notices—E. B. Wilson." *Royal Society Obituary Notice* vol. 3, no. 8, 123–138.
213. MORGAN, T. H., and C. B. BRIDGES. 1913. "Dilution Effects and Bicolorism in Certain Eye Colors of Drosophila." *Journal of Experimental Zoology* 15:429–466.
214. MORGAN, T. H., and C. B. BRIDGES. 1916. "Sexlinked Inheritance in Drosophila." *Carnegie Institute of Washington Publication no. 237.*
215. MORGAN, T. H., C. B. BRIDGES, and A. H. STURTEVANT. 1925. "The Genetics of Drosophila." *Bibliographia Genetica* 2:1–262.
216. MORGAN, T. H., and E. CATTELL. 1912. "Data for the Study of Sexlinked Inheritance in Drosophila." *Journal of Experimental Zoology* 13:79–101.
217. MORGAN, T. H., A. H. STURTEVANT, H. J. MULLER, and C. B. BRIDGES. 1915. *The Mechanism of Mendelian Heredity.* Henry Holt and Co., New York.
218. MULLER, H. J. 1914. "The Bearing of the Selection Experiments of Castle and Phillips on the Variability of Genes." *American Naturalist* 48:567–576.

219. Muller, H. J. 1914. "A Gene for the Fourth Chromosome of Drosophila." *Journal of Experimental Zoology* 17:325–336.
220. Muller, H. J. 1916. "The Mechanism of Crossing Over." *American Naturalist* 50:193–221, 284–305, 350–366, 421–434.
221. Muller, H. J. 1917. "An Oenothera-like Case in Drosophila." *Proceedings of the National Academy of Science* 3:619–626.
222. Muller, H. J. 1918. "Genetic Variability, Twin Hybrids, and Constant Hybrids, in a Case of Balanced Lethal Factors." *Genetics* 3:422–499.
223. Muller, H. J. 1920. "Are the Factors of Heredity Arranged in a Line?" *American Naturalist* 54:97–121.
224. Muller, H. J. 1920. "Further Changes in the White-eye Series in Drosophila and Their Bearing on the Manner of Occurrence of Mutation." *Journal of Experimental Zoology* 31:443–473.
225. Muller, H. J. 1922. "Variation Due to Change in the Individual Gene." *American Naturalist* 56:32–50.
226. Muller, H. J. 1923. "Mutation." *Eugenics, Genetics and the Family* 1:106–112.
227. Muller, H. J. 1926. "The Gene as the Basis of Life." *Proceedings of the International Congress of Plant Science* 1:897–921.
228. Muller, H. J. 1927. "Artificial Transmutation of the Gene." *Science* 66:84–87.
229. Muller, H. J. 1928. "The Production of Mutations by X-rays." *Proceedings of the National Academy of Science* 14:714–726.
230. Muller, H. J. 1928. "The Problem of Genic Modification." *Proceedings of the 5th International Congress of Genetics, Berlin, 1927, Zeitschrift für induktive Abstammungs- und Vererbungslehre* suppl. 1:234–260.
231. Muller, H. J. 1930. "Radiation and Genetics." *American Naturalist* 64:220–225.
232. Muller, H. J. 1930. "Types of Visible Variations Induced by X-rays in Drosophila." *Journal of Genetics* 22:20–334.
233. Muller, H. J. 1932. "Further Studies on the Nature and Causes of Gene Mutations." *Proceedings of the 6th International Congress of Genetics* 1:213–255.
234. Muller, H. J. 1934. "Lenin's Doctrines in Relation to Genetics." In *To the Memory of V. I. Lenin*, pp. 565–592. Moscow-Leningrad Press of Acad. Sci.
235. Muller, H. J. 1935. "A Viable Two-gene Deficiency." *Journal of Heredity* 26:469–478.
236. Muller, H. J. 1935. "The Origination of Chromatin Deficiency as Minute Deletions Subject to Insertion Elsewhere." *Genetica* 17:237–252.
237. Muller, H. J. 1935. "On the Dimensions of Chromosomes and Genes in Dipteran Salivary Glands." *American Naturalist* 69:405–411.
238. Muller, H. J. 1936. "Bar Duplication." *Science* 83:528–530.
239. Muller, H. J. 1938. "The Position Effect as Evidence of the Localization of the Immediate Products of Gene Activity." *Proceedings of the 15th International Physiology Congress, Lengingrad 1935. Secherov J. Physiol. USSR* 21:587–589.
240. Muller, H. J. 1938. "The Present Status of the Mutation Theory." *Current Science*, March, 1938.
241. Muller, H. J. 1940. "An Analysis of the Process of Structural Change in Chromosomes of Drosophila." *Journal of Genetics* 40:1–66.

242. MULLER, H. J. 1943. "Edmund B. Wilson—an Appreciation." *American Naturalist* 77:5–37, 142–172.
243. MULLER, H. J. 1945. "The Gene." *Proceedings of the Royal Society of Biology* 134:1–37.
244. MULLER, H. J. 1954. *Radiation Biology*, edited by A. Hollaender. McGraw-Hill Book Co., New York.
245. MULLER, H. J. 1956. "On the Relation between Chromosome Changes and Gene Mutations." *Brookhaven Symposia in Biology* 8:126–147.
246. MULLER, H. J., and E. ALTENBURG. 1919. "The Rate of Change of Hereditary Factors in Drosophila." *Proceedings of the Society of Experimental Biology and Medicine* 17:10–14.
247. MULLER, H. J., and E. ALTENBURG. 1930. "The Frequency of Translocations Produced by X-rays in Drosophila." *Genetics* 15:283–311.
248. MULLER, H. J., and A. PROKOFYEVA. 1934. "Continuity and Discontinuity of the Hereditary Material." *C. R. (Doklady) de l'Acad. des Sci. USSR* vol. 4 no. 1, pp. 8–12.
249. MULLER, H. J., and A. A. PROKOFYEVA. 1935. "The Individual Gene in Relation to the Chromomere and the Chromosome." *Proceedings of the National Academy of Science* 21:16–26.
250. MULLER, H. J., A. A. PROKOFYEVA-BELGOVSKAYA, and K. V. KOSSIKOV. 1936. "Unequal Crossing-over in the Bar Mutant as a Result of Duplication of a Minute Chromosome Section." *C. R. (Doklady) de l'Acad. des Sci. USSR* 1(k), #2 (79), pp. 87–88.
251. MULLER, H. J., A. PROKOFYEVA, and D. RAFFEL. 1935. "Minute Intergenic Rearrangement as a Cause of Apparent 'Gene Mutation.' " *Nature* 135:253–255.
252. NANNEY, D. L. 1957. "The Role of the Cytoplasm in Heredity." In *The Chemical Basis of Heredity*, edited by W. D. McElroy and H. B. Glass. Johns Hopkins Press, Baltimore, Md.
253. NIRENBERG, M. W., and J. H. MATTHAEI. 1961. "The Dependence of Cell-free Protein Synthesis in E. coli upon Naturally Occurring or Synthetic Polyribonucleotides." *Proceedings of the National Academy of Science* 47:1588–1594.
254. NOVICK, A., and M. WEINER. 1957. "Enzyme Induction as an All-or-None Phenomenon." *Proceedings of the National Academy of Science* 43:553–566.
255. OFFERMANN, C. A. 1935. "The Position Effect and Its Bearing on Genetics." *Izvestia Akademii Nauk. SSSR 1935, Bull. de l'Acad. des Sci. de l'URSS* 7th series, pp. 129–140.
256. OLIVER, C. P. 1940. "A Reversion to Wild-type Associated with Crossing-over in Drosophila Melanogaster." *Proceedings of the National Academy of Science* 26:452–454.
257. PAULING, L., H. A. ITANO, S. J. SINGER, and I. C. WELLS. 1949. "Sickle Cell Anemia, a Molecular Disease." *Science* 110:543.
258. PEARSON, E. S. 1938. *Karl Pearson: Some Aspects of His Life and Work.* Cambridge University Press, Cambridge.
259. PONTECORVO, G. 1952. "Genetic Formulation of Gene Structure and Gene Action." *Advances Enzym.* 13:121–149.
260. PONTECORVO, G. 1955. "Gene Structure and Action in Relation to Heterosis." *Proceedings of the Royal Society of Biology* 144:171–177.
261. PONTECORVO, G. 1956. "Allelism." *Cold Spring Harbor Symposia on Quantitative Biology* 21:171–174.

262. PONTECORVO, G. 1958. *Trends in Genetic Analysis.* Columbia University Press, New York.

263. PONTECORVO, G., and J. A. ROPER. 1956. "Resolving Power of Genetic Analysis." *Nature* 178:83–84.

264. PREER, J. R. 1946. "Some Properties of a Genetic Cytoplasmic Factor in P." *Proceedings of the National Academy of Science* 32:247–253.

265. PREER, J. 1948. "The Killer Cytoplasmic Factor Kappa: Its Rate of Reproduction, the Number of Particles per Cell." *American Naturalist* 82:35–42.

266. QUACKENBUSH, L. S. 1910. "Unisexual Broods of Drosophila." *Science* 32:183–185.

267. RAFFEL, D., and H. J. MULLER. 1940. "Position Effect and Gene Divisibility Considered in Connection with Three Strikingly Similar Scute Mutations." *Genetics* 25:541–583.

268. RHOADES, M. M. 1943. "Genic Induction of an Inherited Cytoplasmic Difference." *Proceedings of the National Academy of Science* 29: 327–329.

269. RHOADES, M. M. 1946. "Plastid Mutations." *Cold Spring Harbor Symposia on Quantitative Biology* 11:202–207.

270. ROPER, J. A. 1950. "A Search for Linkage between Genes Determining Vitamin Requirements." *Nature* 166:956.

271. SAFIR, S. R. 1913. "A New Eye Color Mutation in Drosophila and Its Mode of Inheritance." *Biology Bulletin* 25:45–51.

272. SAFIR, S. R. 1916. "Buff, a New Allelomorph of White Eye Color in Drosophila." *Genetics* 1:584–590.

273. SCHRÖDINGER, E. 1945. *What is Life?* Cambridge (The Macmillan Co., New York).

274. SEREBROVSKY, A. S. 1927. "The Influence of the 'Purple' Gene on the Crossing-over between 'Black' and 'Cinnabar' in Drosophila Melanogaster." *Journal of Genetics* 18:137–175.

275. SEREBROVSKY, A. S., O. A. IVANOVA, and L. FERRY. 1929. "On the Influence of Genes y, l¹, N¹ on the Crossing Over Close to Their Loci in the Sex-chromosome of Drosophila Melanogaster." *Journal of Genetics* 21:287–314.

276. SHULL, G. H. 1909. "The Presence and Absence Hypothesis." *American Naturalist* 43:410–419.

277. SHULL, G. H. 1912. " 'Genes' or 'Gens'?" *Science,* n. s., 35:819.

278. SHULL, G. H. 1935. "The Word 'Allele.' " *Science* 82:37–38.

279. SONNEBORN, T. M. 1937. "Sex, Sex Inheritance and Sex Determination in Paramecium aurelia." *Proceedings of the National Academy of Science* 23:378–385.

280. SONNEBORN, T. M. 1939. "Sexuality and Related Problems in Paramecium." *Collecting Net* 14:1–6.

281. SONNEBORN, T. M. 1943. "Gene and Cytoplasm. I. The Determination and Inheritance of the Killer Character in Variety 4 of Paramecium Aurelia." *Proceedings of the National Academy of Science* 29:329–343.

282. SONNEBORN, T. M. 1943. "Gene and Cytoplasm. II. The Bearing of the Determination and Inheritance of Characters in Paramecium Aurelia on the Problems of Cytoplasmic Inheritance, Pneumococcus Transformations, Mutations and Development." *Proceedings of the National Academy of Science* 29:338–343.

283. SONNEBORN, T. M. 1946. "Experimental Control of the Concentration

of Cytoplasmic Factors in Paramecium." *Cold Spring Harbor Symposia on Quantitative Biology* 11:236–255.

284. SONNEBORN, T. M. 1947. "Developmental Mechanisms in Paramecium." *Growth Symposium* 11:291–307.

285. SONNEBORN, T. M. 1948. "Symposium on Plasmagenes, Genes, and Characters in Paramecium Aurelia," Introduction. *American Naturalist* 82:26–34.

286. SONNEBORN, T. M. 1949. "Beyond the Gene." *American Scientist* 37: 33–59.

287. SONNEBORN, T. M. 1950. "The Cytoplasm in Heredity." *Heredity* 4: 11–36.

288. SONNEBORN, T. M. 1955. "Heredity, Development, and Evolution in Paramecium." *Nature* 175:1100.

289. SONNEBORN, T. M. 1959. "Kappa and Related Particles in Paramecium." *Advances in Virus Research* 1959:229–356.

290. SONNEBORN, T. M. 1961. "Kappa Particles and Their Bearing on Host-parasite Relations." *Perspectives in Virology* 2:5–12.

291. STADLER, L. J. 1928. "Mutations in Barley Induced by X-rays and Radium." *Science* 68:186–187.

292. STADLER, L. J. 1932. "On the Genetic Nature of Induced Mutations in Plants." *Proceedings of the 6th International Congress of Genetics* 1:274–294.

293. STADLER, L. J. 1941. "The Comparison of Ultraviolet and X-ray Effects on Mutation." *Cold Spring Harbor Symposia on Quantitative Biology* 9:108–177.

294. STADLER, L. J. 1954. "The Gene." *Science* 120:811–819.

295. STURTEVANT, A. H. 1913. "The Himalayan Rabbit Case, with Some Considerations on Multiple Allelomorphs." *American Naturalist* 47:234–238.

296. STURTEVANT, A. H. 1913. "A Third Group of Linked Gene in Drosophila Ampelaphila." *Science* 37:990–992.

297. STURTEVANT, A. H. 1913. "The Linear Arrangement of Six Sex-linked Factors in Drosophila, as Shown by Their Mode of Association." *Journal of Experimental Zoology* 14:43–59.

298. STURTEVANT, A. H. 1914. "The Reduplication Hypothesis as Applied to Drosophila." *American Naturalist* 48:535–549.

299. STURTEVANT, A. H. 1915. "The Behavior of the Chromosomes as Studied through Linkage." *Zeitschrift für induktive Abstammungs- und Verersbungslehre* 13:234–287.

300. STURTEVANT, A. H. 1917. "Crossing Over without Chiasmatype?" *Genetics* 2:301–304.

301. STURTEVANT, A. H. 1925. "The Effects of Unequal Crossing Over at the Bar Locus in Drosophila." *Genetics* 10:117–147.

302. STURTEVANT, A. H. 1926. "A Crossover Reduced in Drosophila Melanogaster Due to Inversion of a Section of the Third Chromosome." *Biologischen Zentralblatt* 46:697–702.

303. STURTEVANT, A. H. 1928. "A Further Study of the So-called Mutation at the Bar Locus of Drosophila." *Genetics* 13:401–409.

304. STURTEVANT, A. H. 1959. "Thomas Hunt Morgan." *Biographical Memoirs* 33:283–325.

305. STURTEVANT, A. H., and T. H. MORGAN. 1923. "Reverse Mutation of the Bar Gene Correlated with Crossing Over." *Science* 57:746–747.

306. STURTEVANT, A. H., and J. SCHULTZ. 1931. "The Inadequacy of the Subgene Hypothesis of the Nature of the Scute Allelomorphs of Dro-

sophila." *Proceedings of the National Academy of Science* 71:265–270.

307. THOMPSON, D. H. 1931. "The Side Chain Theory of the Structure of the Gene." *Genetics* 16:267–290.

308. TICE, S. C. 1914. "A New Sex-linked Character in Drosophila." *Biological Bulletin* 26:221–230.

309. TIMOFEEF-RESSOVSKY, N. W., E. G. ZIMMER, and M. DELBRÜCK. 1935. "Über die Natur der Genmutation und der Genstrucktur." *Nachr. a. d. Biologie d. Ges. d. Wiss Göttingen* 1:189.

310. TROLAND, L. T. 1917. "Biological Enigmas and the Theory of Enzyme Action." *American Naturalist* 51:321–350.

311. TROW, A. H. 1913. "Forms of Reduplication—Primary and Secondary." *Journal of Genetics* 2:313–324.

312. VOGEL, H. J. 1957. "Repressed and Induced Enzyme Formation: a Unified Hypothesis." *Proceedings of the National Academy of Science* 43:491–496.

313. VOLKIN, E., and L. ASTRACHAN. 1956. "Phosphorus Incorporation in Escherichia Coli Ribonucleic Acid after Infection with Bacteriophage T2." *Virology* 2:149–161.

314. WAHBA, A. J., R. S. MILLER, C. BASILIO, R. S. GARDNER, P. LENGYEL, and J. F. SPOYER. 1963. "Synthesis of Polynucleotides and the a. a. Code IX." *Proceedings of the National Academy of Science* 49:880–885.

315. WATSON, J. D., and F. H. C. CRICK. 1953. "Molecular Structure of Nucleic Acids." *Nature* 171:737–738.

316. WATSON, J. D., and F. H. C. CRICK. 1953. "Genetical Implications of the Structure of Deoxyribonucleic Acid." *Nature* 171:964.

317. WELDON, W. F. R. 1898. "Presidential Address, Zoological Section (D)." *British Association for the Advancement of Science*, Bristol, 1898, pp. 1–16.

318. WELDON, W. F. R. 1901. "Mendel's Laws of Alternative Inheritance in Peas." *Biometrika* 1:228–254.

319. WILSON, E. B. 1896. *The Cell*. The Macmillan Co., New York.

320. WOODWARD, D. O. 1959. "Enzyme Complementation in vitro between Adenylosuccinase Mutants of Neurospora Crassa." *Proceedings of the National Academy of Science* 45:846.

321. YANOFSKY, C., B. C. CARLTON, J. R. GUEST, D. R. HELINSKI, and U. HENNING. 1964. "On the Colinearity of Gene Structure and Protein Structure." *Proceedings of the National Academy of Science* 51:266–272.

322. ZELENY, C. 1921. "The Direction and Frequency of Mutation in the Bar Eye Series of Multiple Allelomorphs of Drosophila." *Journal of Experimental Zoology* 34:203–233.

323. ZINDER, N., and J. LEDERBERG. 1952. "Genetic Exchange in *Salmonella*." *Journal of Bacteriology* 64:679–699.

Index of Author References

289

Index of Subjects